Longitudinal Study of ITS Implementation: Decision Factors and Effects

Final Report

www.its.dot.gov/index.htm
Final Report — April 2013
Publication Number: FHWA-JPO-13-067

U.S. Department of Transportation
Research and Innovative Technology Administration

Produced by Noblis, Inc.
U.S. Department of Transportation
ITS Joint Program Office
Research and Innovative Technology Administration

Notice

This document is disseminated under the sponsorship of the Department of Transportation in the interest of information exchange. The United States Government assumes no liability for its contents or use thereof.

The U.S. Government is not endorsing any manufacturers, products, or services cited herein and any trade name that may appear in the work has been included only because it is essential to the contents of the work.

Cover Photo Credit:
Top row: iStockphoto/RiverNorthPhotography 2013; iStockphoto/WendellandCarolyn 2013; Think Stock 2013
Middle row: USDOT 2003, Noblis 2012
Bottom row: Noblis, 2012; Think Stock/Comstock 2013; iStockphoto/Slobo 2013

Technical Report Documentation Page

1. Report No. FHWA-JPO-13-067	2. Government Accession No.	3. Recipient's Catalog No.
4. Title and Subtitle Longitudinal Study of ITS Implementation: Decision Factors and Effects Final Report		5. Report Date
		6. Performing Organization Code
7. Author(s) Vaishali Shah, Carolina Burnier, Drennan Hicks, Greg Hatcher, Liz Greer, Doug Sallman, William Ball, Katie Fender, Dan Murray		8. Performing Organization Report No.
9. Performing Organization Name And Address Noblis 600 Maryland Ave., SW, Suite 755 Washington, DC 20024		10. Work Unit No. (TRAIS)
		11. Contract or Grant No. DTFH61-11-D-00018
12. Sponsoring Agency Name and Address ITS-Joint Program Office Research and Innovative Technology Administration 1200 New Jersey Avenue, S.E. Washington, DC 20590		13. Type of Report and Period Covered Final Report
		14. Sponsoring Agency Code HOIT-1

15. Supplementary Notes
James Pol, COTR

16. Abstract

The Intelligent Transportation Systems (ITS) Joint Program Office (JPO) is placing increasing emphasis on transferring ITS technology from research to deployment, and on accelerating the rate of ITS technology adoption. As part of these efforts, the JPO has sponsored research studies intended to improve the state of knowledge regarding the underlying characteristics and factors for technology adoption and deployment. This report is the final deliverable from the most recent of these studies, the Longitudinal Study of Implementation: Decision Factors and Effects (started in January 2012). This final report documents the findings and key observations from all tasks of the Longitudinal Study of Implementation.

The Longitudinal Study of Implementation builds upon a body of existing work related to decision factors influencing ITS adoption, growth, maintenance or decline within the public and private sectors. The Longitudinal Study uses an interview-based approach to further analyze decision factors among public sector transportation agencies and the trucking industry; interviews with connected vehicle technology representatives from the automotive industry to assess their perspectives on what is needed for the connected vehicle environment to be fully realized; a post-hoc set of studies reviewing deployments, costs, and benefits at early ITS deployment sites; and a workshop and analysis of how to present cost and benefit information in a way that best informs and influences decision-makers. Finally, based on a cross-cutting assessment of these findings, the study team suggests several major themes for the federal government to consider regarding next generation ITS and the connected vehicle environment.

Results indicate that for the public sector, the most important technology and application factor was quality and reliability, followed by interoperability considerations and demonstration of benefits. The most important external factor was budget and funding sources. For the trucking industry, the most important factors for adopting a new technology were the price/ Return-on-Investment (ROI), compatibility with existing systems, readiness and maturity of the technology, quality and reliability, and product service and support.

17. Key Words Intelligent Transportation Systems, factors influencing ITS adoption and deployment, technology diffusion, innovation theory, longitudinal study of implementation, post-hoc data analysis, connected vehicle.	18. Distribution Statement		
19. Security Classif. (of this report) Unclassified	20. Security Classif. (of this page) Unclassified	21. No. of Pages 78	22. Price

Form DOT F 1700.7 (8-72)   Reproduction of completed page authorized

Preface/Acknowledgements

The Intelligent Transportation Systems (ITS) Joint Program Office (JPO) is placing increasing emphasis on transferring ITS technology from research to deployment, and on accelerating the rate of ITS technology adoption. The JPO technology transfer efforts are aimed at accelerating deployment of both current and near-future technologies, such as connected vehicle. As part of these efforts, the JPO has sponsored research studies intended to improve the state of knowledge regarding the underlying characteristics and factors for technology adoption and deployment. This report is the final deliverable from the most recent of these studies, the Longitudinal Study of Implementation: Decision Factors and Effects (started in January 2012). This final report documents the findings and key observations from all tasks of the Longitudinal Study of Implementation.

The Longitudinal Study of Implementation builds upon a body of existing work related to decision factors influencing ITS adoptions, growth, maintenance or decline within the public and private sectors. The Longitudinal Study uses an interview-based approach to further analyze decision factors among public sector transportation agencies and the trucking industry; interviews with connected vehicle technology representatives from the automotive industry to assess their perspectives on what is needed for the connected vehicle environment to be fully realized; a set of studies reviewing deployments, costs, and benefits at early ITS deployment sites; and a workshop and analysis of how to present cost and benefit information in a way that best informs and influences decision-makers. Finally, based on a cross-cutting assessment of these findings, the research team suggests several major themes for the federal government to consider regarding next generation ITS and the connected vehicle environment.

Table of Contents

Executive Summary 1
    STUDY OBJECTIVES 1
    DECISION FACTORS FOR IMPLEMENTATION OF ITS 1
    VALUATION OF BENEFITS AFTER INITIAL ITS DEPLOYMENT 3
    SUPPORT FOR ITS INVESTMENT DECISIONS 4
    CONSIDERATIONS FOR NEXT GENERATION ITS AND CONNECTED VEHICLE 5

1   Introduction 8
2   Methodology and Process 9
    2.1 LITERATURE REVIEW 9
        2.1.1   WEBINAR RESULTS 10
    2.2 STAKEHOLDER INTERVIEWS 11
    2.3 POST-HOC ANALYSIS 13
    2.4 PUBLIC WORKSHOP 13

3   Public Sector Perspective 14
    3.1 IMPORTANCE OF ITS DECISION FACTORS 15
    3.2 ITS IMPLEMENTATION BARRIERS 19
    3.3 SUPPORT FOR ITS INVESTMENT DECISIONS 21
        3.3.1   EVOLUTION OF THE ITS DECISION MAKING PROCESS 21
        3.3.2   ROLE OF PERFORMANCE MEASUREMENT IN ITS 22
        3.3.3   PRIORITIZATION IN A CONSTRAINED ECONOMIC ENVIRONMENT 23
        3.3.4   INTEROPERABILITY AND INTEGRATION 24
        3.3.5   AWARENESS AND IMPLICATIONS OF CONNECTED VEHICLE TECHNOLOGY 24

4   Trucking Industry Perspective 26
5   Automotive Manufacturer Perspective 29
    5.1 U.S. DOT CONNECTED VEHICLE RESEARCH PROGRAM 29
    5.2 CONSIDERATIONS FOR MOVING FORWARD TO A CONNECTED VEHICLE ENVIRONMENT 31

6   Post-Hoc Analyses 33
7   Public Workshop Summary 37
    7.1 WORKSHOP OBJECTIVES 37
    7.2 WORD CLOUD 37
    7.3 KEY THEMES 38
    7.4 PARTICIPANT RESPONSE 40

8   Considerations for Next Generation ITS 41
    8.1 APPLYING PAST EXPERIENCE TO ACHIEVE FUTURE SUCCESS 42
    8.2 CROSS-CUTTING THEMES 42
    8.3 AREAS FOR FUTURE RESEARCH 44

References 45
APPENDIX A.   List of Project Reports 46
APPENDIX B.   List of Acronyms 47
APPENDIX C.   Site Visit Summaries 50

  GEORGIA 50
  IDAHO 52
  MARYLAND 54
  PHOENIX, ARIZONA 56
  TUCSON, ARIZONA 57
  WASHINGTON STATE AND NEW JERSEY 59

APPENDIX D.   Interview Instruments 61
  PUBLIC SECTOR PRE/POST WEBINAR SURVEY 61
  PUBLIC SECTOR SCREENING INTERVIEW FORM 62
  PUBLIC SECTOR IN-DEPTH INTERVIEW FORM 65
  TRUCKING INDUSTRY INTERVIEW FORM 70

APPENDIX E.   Webinar Summary 74
  WEBINAR OBJECTIVES AND AUDIENCE 74
  FACTORS INFLUENCING ITS ADOPTION 75
  FACTORS INFLUENCING DECISION TO GROW, MAINTAIN, CONTRACT/CANCEL ITS 76
  WEBINAR CONCLUSION AND FOLLOW UP 77

List of Tables

Table 3-1: Detailed Importance Rating Statistics for Factors Influencing the Decision to Implement ITS *(Source: Noblis 2013)* 17

Table 8-1: Stakeholder Top Decision Factors and Key Questions regarding Connected Vehicle Technology *(Source: Noblis 2013)* 41

List of Figures

Figure ES-1: Sets of Factors Influencing ITS Implementation Decisions *(Source: Noblis 2013)* 2

Figure 2-1: Longitudinal Study of ITS Implementation (LSI) Key Tasks *(Source: Noblis 2013)* 9

Figure 2-2: Activities Implemented and Organization Types Interviewed to Address Concepts, Factors, and Issues Related to ITS Decision-Making *(Source: Noblis 2013)* 12

Figure 3-1: Site visit to Boise, Idaho *(Source: Noblis 2012)* 14

Figure 3-2: Public Sector Visit Locations *(Source: Noblis 2013)* 14

Figure 3-3: Sets of Factors Influencing ITS Implementation Decisions *(Source: Noblis 2013)* 15

Figure 3-4: Three Stages of ITS Decision Making *(Source: Noblis 2013)* 16

Figure 3-5: Most Important Factors during Initiation Phase of ITS Implementation *(Source: Noblis 2013)* 18

Figure 3-6: Most important Factors during Development Phase of ITS Implementation *(Source: Noblis 2013)* 18

Figure 3-7: Most Important Factors during Deployment Phase of ITS Implementation *(Source: Noblis 2013)* 19

Figure 3-8 Rating of Importance of Barriers to ITS Implementation *(Source: Noblis 2013)* ..20

Figure 3-9: Traffic Management Center in Boise, ID *(Source: Noblis 2012)* 21

Figure 3-10: Maryland DOT Traffic Management Center *(Source: Noblis 2012)* 22

Figure 3-11: Atlanta, Georgia HOT lanes *(Source: Noblis 2012)* 23

Figure 3-12: Northwest Passage *(Courtesy of Northwest Passage Pooled Fund Study 2013)* 24

Figure 3-13: Connected vehicle environment *(Source: U.S. DOT 2013)* 25

Figure 4-1: Number of decision makers based on fleet size *(Source: ATRI 2013)* 27

Figure 4-2: Connected freight vehicle *(Source: U.S. DOT 2013)* 28

Figure 5-1: Crash Avoidance Metrics Partnership Vehicle Safety Communications Consortium (VSCC) members *(Source: National Highway Traffic Safety Administration 2013)* 29

Figure 5-2: Connected vehicle environment *(Source: U.S. DOT 2013)* 30

Figure 5-3: Examples of connected vehicle competitive technologies *(Source from left to right: Think Stock 2013; courtesy of Volvo 2013; Steve Jurvetson/Wikimedia Commons/Public Domain; derivative work: Mariordo. 2013)* 31

Figure 6-1: Traveler Information via mobile device *(Source: Think Stock 2013)* 34

Figure 6-2: Ramp metering *(Source: Wikimedia Commons/Patriarca12 2013)* 34

Figure 6-3: MnPass HOT Lanes Pass *(Courtesy Minnesota DOT 2013)* 35

Figure 6-4: Traffic Signal Control *(Source: Think Stock/David De Lossy 2013)* 35

Figure 7-1: Connected Vehicle Word Cloud *(Source: Noblis 2013)* 38

Figure 7-2: Interactive Exercise - Connected Vehicle Word Wall *(Source: Noblis 2013)* 40

Figure C-1: Transportation Management Centers in Georgia *(Source: Noblis 2012)* 50

Figure C-2: Transportation Agencies in Idaho *(Source: Noblis 2012)* 52
Figure C-3: Maryland Coordinated Highway Action Response Team (CHART) Facilities
    (Source: Noblis 2012) 54
Figure C-4: Transportation Agencies in Phoenix, Arizona *(Source: Noblis 2012)*...................56
Figure C-5: Traffic Management Centers in Tucson, AZ *(Source: Noblis 2012)*....................57
Figure E-6: Webinar participant affiliation *(Source: Noblis 2013)* 74
Figure E-7: Factors Cited as Most Important When Deciding to Adopt an ITS Technology or
    System *(Source: Noblis 2013)* 75
Figure E-8: Factors Selected As Most Important during Decisions to Expand, Maintain, or
    Contract/Cancel ITS Technologies or Systems *(Source: Noblis 2013)*..........................77

Executive Summary

With almost 20 years of ITS deployment experience behind us, Intelligent Transportation Systems (ITS) is at a crossroads, with the first generation of ITS technologies at a saturation point for mature ITS applications, especially in the large metropolitan areas across the United States. As we move forward toward the connected vehicle environment and coordinated operations system envisioned for the future, understanding the motivating factors used by state and local agencies, automobile manufacturers, and the commercial vehicle industry for adopting ITS technology is of critical importance.

The Longitudinal Study of ITS implementation (LSI), conducted by Noblis and its partners, American Transportation Research Institute (ATRI), Merriweather Advisors, and Cambridge Systematics, provides a foundation that captures the state of knowledge for motivating factors influencing ITS adoption, maintenance, and for continuing its use and deployment through:

- a comprehensive literature review on technology innovation;
- an interview-based approach to further analyze decision factors;
- post-hoc studies reviewing deployments, costs, and benefits at early ITS deployment sites;
- and a workshop focusing on considerations for future ITS policy and initiatives.

Study Objectives

The objectives of the study were to answer the following research questions:

- What are the most important factors that influence adoption, growth, replacement or cancellation of ITS technology? Do these factors differ according to ITS implementation phase?
- Does expansion of an ITS deployment produce a higher level of benefits? Do benefits trail off as the ITS deployment is operated for a long period?
- What information and delivery methods best support stakeholders in making ITS planning, implementation, O&M, and replacement decisions?
- What can the U.S. DOT do to accelerate ITS technology implementation? What is the applicability of these strategies to connected vehicle advances and next generation ITS?

The answers to these questions are discussed in the following subsections.

Decision Factors for Implementation of ITS

The Noblis team analyzed information obtained from public sector, trucking, and automobile manufacturer decision makers to address the key factors and issues related to decisions concerning the adoption, growth, maintenance, replacement or cancellation of ITS systems. The locations chosen for site visits represented diversity according to geography, metropolitan size, agency responsibility (transit, highway or arterial), level of congestion, ITS decision time frame, and maturity of technology deployed. The set of decision factors were organized into four categories: technology or application factors, implementer factors, external environment factors, and user/market factors, as presented in Figure ES-1. The relative importance of each of these categories and specific factors were explored qualitatively and quantitatively through the public sector interviews. These factors were modified and assessed with the trucking industry to identify factors of importance this sector.

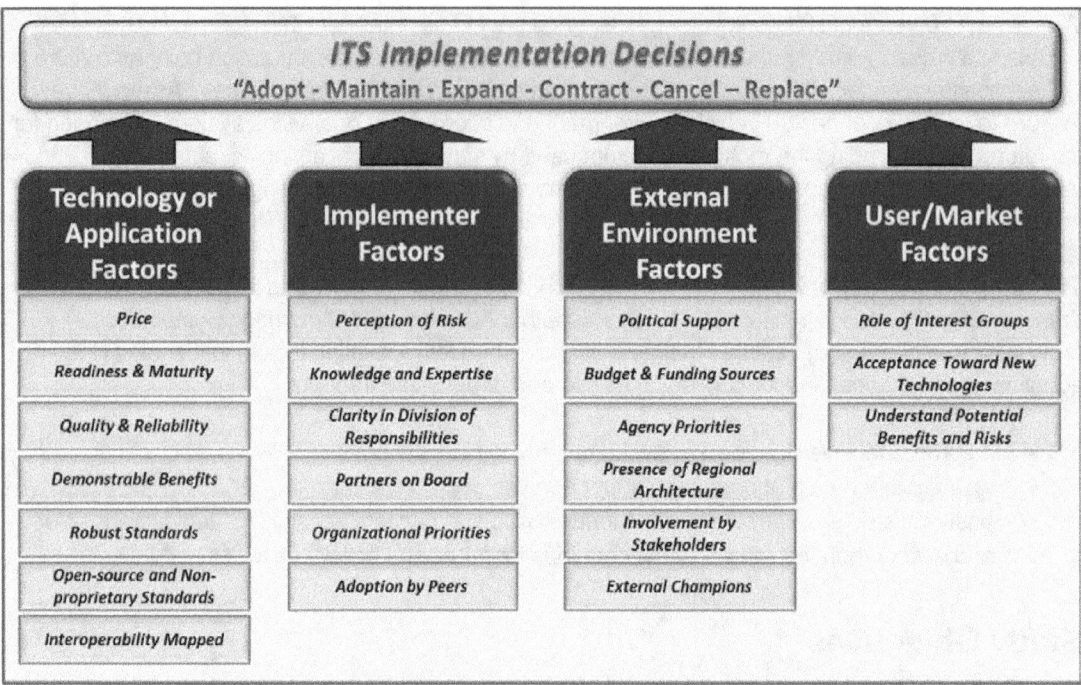

Figure ES-1: Sets of Factors Influencing ITS Implementation Decisions *(Source: Noblis 2013)*

Across the three phases of ITS implementation, the most important technology and application factor was **quality and reliability**, followed by **interoperability** considerations and **demonstration of benefits**. The most important among implementing organization factors are **having partners on board** and **organizational priorities**. The most important external factor was clearly **budget and funding** sources. This factor was also the highest rated among all factors on average across the three phases of ITS implementation. These factors were on average important across all regions interviewed.

Importance of Factors by Phase of ITS Implementation
One of the gaps identified by the literature review was that motiving factors may change according to the phase of ITS implementation. The study asked public sector stakeholders to rank the decision factors by phase of implementation: initiation, development or deployment. During the initiation phase, budget and funding was most critical (rating of 7.8). Demonstration of benefits and involvement in the project by stakeholders comes in second, both with an average rating of 7.2. Moving to the development phase, budget and funding continued to be important with an even higher rating, that of 8.4 out of 10. During the development phase, interoperability came in second, with an average rating of 7.9. In the deployment phase, quality and reliability (8.9 rating) and end-user awareness/understanding (8.4 rating) superseded budget and funding (8.3 rating). During this phase of ITS implementation, a number of organizational factors took precedence including staff knowledge and expertise, clarity in division of responsibilities, and having partners on board (all with 8.0 rating).

Executive Summary

These findings were confirmed by qualitative analysis of the interviews. For adoption and growth decisions that discussed technology and application factors, the largest percentage of interviewees cited demonstration of benefits (55%) as a factor in the decision to adopt or grow the technology. This percentage was smaller for other types of decisions, such as replace (27%), cancel (20%) and not select a technology (6%). Across the board, for these other types of decisions, quality and reliability of technology was cited most often as the technology factor.

Barriers to ITS Adoption
The respondents were also asked to rate barriers to ITS adoption. As the ITS implementation moves from initiation, through development, and into deployment, knowledge, technical, and societal barriers became more critical. Legal and regulatory, financial and economic, and decision making barriers were more important during the development stage compared to the initiation or deployment stages of ITS implementation.

Trucking Industry Factors
For the trucking industry, the key factors considered before investing in technology were:
- Price of the technology/return-on-investment
- Readiness and maturity of the technology
- Compatibility with the existing systems
- System integration and flexibility
- Quality, reliability, service and support

As expected, the factors demonstrate that the trucking industry places considerable importance on demonstration of monetary benefits and less emphasis on inter-agency coordination and cooperation.

Looking overall at factors influencing the implementation of ITS, budget/funding, quality and reliability, demonstration of benefits, and compatibility/standards are most important. End-user awareness and acceptance were also rated as important factors for ITS implementation. A knowledgeable and skilled workforce, budget/funding, and compatibility & standards challenges were cited as critical barriers to ITS implementation. It has been noted that many of these factors are expressed in the framework for the Capability and Maturity Model developed for Transportation System Operations & Management (SO&M). The six dimensions of that framework (Business Processes, Systems & Technology, Performance Measurement, Culture, Organization/Workforce, and Collaboration) are used to provide a structured approach to perform self-assessment and identify the incremental changes in agency capabilities that are essential to improving SO&M effectiveness.

Valuation of Benefits after Initial ITS Deployment

The post-hoc analyses examined how the performance of various systems have changed over time, either due to expansion/enhancement of the systems, or changing traffic patterns or traveler behavior. Evaluations include assessment of:

- a transit traveler information system in Portland, Oregon
- a ramp metering deployment in Kansas City,
- high occupancy toll (HOT) lanes in Minneapolis/St. Paul, and
- an arterial management system in Phoenix, Arizona.

Executive Summary

Three of these analyses were carried out in time horizons with fairly stable, built-out network environments, and for these analyses, results show meaningful trends or implications as described in chapter 7. The single evaluation in Phoenix, Arizona with significant variations in population and demand during the evaluation period proved more complex to evaluate and interpret.

Overall, the post-hoc evaluations demonstrated:

1. Ongoing and expanded ITS implementations continue to produce measurable effects. In some cases, initial benefits can be sustained over time. In other cases, the benefits may change or decrease as the system infrastructure and operational strategies evolve, but the implementation can still produce a positive benefit-cost ratio, especially if the incremental change in operational costs decline.
2. Significant advances in technology, such as Portland's transit traveler information system which provided 511 transit arrival times over mobile devices, can have a disruptive effect on benefits.
3. Archived performance data from areas with sufficient data quality procedures can be used to support a data analysis to review trends over time, without the need to collect more field data.

These post-hoc evaluations provide confidence that continued O&M on ITS deployments is a prudent use of public funds. In the future, agencies should factor in the cost side when comparing potential investments as incremental costs are often reduced as system expands. This innovative use of archived performance data offers a glimpse into performance-based management, where agencies will be expected to monitor yearly performance. Agencies can look to the archived data as sources for baseline metrics and measures.

Support for ITS Investment Decisions

The site visits enabled interviewees to expand on five timely topics, each with implications for how the U.S. DOT could best support public sector stakeholders in making ITS planning, implementation, O&M, and replacement decisions. These topics, summarized below, are described in detail in chapter 3 of this Final Report.

Evolution of ITS decision making. There has been a shift from adoption of ITS to operations and maintenance as ITS is now mainstreamed. The complexity of ITS decision-making has increased as the number of decision-makers from partner agencies that an agency must coordinate with when implementing ITS projects has typically grown. With a greater attention on operations and maintenance, there is a clear need for education and training for ITS staff on operations of ITS. The influence of peers is more significant than in the past, but the definition of who serves as a peer varies depending on the region and reflects size and congestion levels. Stakeholders requested that future demonstrations include diverse constituents to ensure representation of all regional characteristics.

Role of performance-based management. With the new legislative provisions for performance-based management and reporting through MAP-21, states agencies are apprehensive of what performance measures will be required and the impact on their funding. State agencies expressed that they are looking for federal guidance for better, clearer information on what performance metrics can most effectively be measured to manage by performance. They also expressed concern about the cost of performance monitoring and noted the difficulty of measuring the performance of some essential ITS technologies such as dynamic message signs, cameras, and training for personnel. Federal guidance on metrics and methodologies were requested in these areas.

Funding Competition in constrained economy. Decision makers understand the value of ITS solutions in a constrained economic environment; however state agencies expressed the challenge of funding these solutions when elected officials are eager to deliver high profile construction projects and ribbon cuttings. The investment in ITS also represents a lifelong investment in operations and maintenance. Benefit-cost analysis showing the sustained benefits of these systems could be helpful in assuring public commitment to these systems.

Several agencies expressed that federal requirements are making it difficult for local jurisdictions to apply for federal funding, and state agencies have also expressed concerns involving using federal funding because of the limits on how those dollars can be spent. The need to share resources across institutions and regions was a common theme. Stakeholders are looking for adjustments in federal funding requirements to facilitate this resource sharing.

Importance of interoperability and integration. Compatibility and interoperability become more critical as ITS installed base increases. A desire for interoperable systems has created new cross jurisdictional and institutional issues that need to be addressed. However, Information Technology (IT) departments have been reluctant to open up internal networks to other agencies due to security concerns. In addition, funding limitations may lead to agencies deploying systems that are not interoperable. Stakeholders cited support for a strong regional organization that brings together stakeholder agencies as tremendously effective in promoting a coordinated adoption, maintenance and growth of ITS.

Connected Vehicle perspectives. Many organizations reported that they are unclear about the objectives of the connected vehicle research program in terms of the proposed applications, particularly in the vehicle-to-infrastructure side, and what the implementation period for these applications would be. Some agencies reported that they do not know how to become involved in the program and feel like they may have "missed the boat." These agencies are looking for opportunities and federal guidance for how to become involved in the program. Decision makers also want more information on the business cases to support justification. There is a concern that connected vehicle applications would be mandated without federal funding to help implement them. Stakeholders expressed that the federal government needs to provide more demonstrations, training, and direction to local and state DOTs.

Considerations for Next Generation ITS and Connected Vehicle

Overall, cross-cutting analysis of the stakeholder interviewers, post-hoc data analysis, and workshop revealed several major themes for the federal government to consider regarding next generation ITS and the connected vehicle environment:

1. ***Clearly define and publicize benefits for connected vehicle technology to engage stakeholder interest.*** Many public organizations are unclear of what the connected vehicle program is and what the benefit of implementation would be. In particular, they are questioning the business model of implementing these new systems. It was clear that demonstrating societal benefits alone may not be sufficient to convince state and local agencies of the value of connected vehicle technology, and that operational efficiencies would be important to these stakeholders.

2. ***Recognize that private sector prefers a market driven approach on the vehicle side while public sector seeks stronger federal guidance on infrastructure deployment.*** The private sector feels that a market based approach should drive connected vehicle implementation, while public agencies are asking for guidance from the Federal government with respect to applications that they would implement, their role in the connected vehicle research program, and what standards would need to ensure interoperability and federal funding. There is a concern that connected vehicle applications will be mandated without federal funding to back them up.

3. ***Ensure demonstrations include diverse constituents.*** Demonstrations of connected vehicle technologies should include diverse constituents in terms of modality, levels of congestion, and size of deployment to establish a robust peer group for market share growth. Some public stakeholders do not know how to get involved and feel like they have missed the boat. For example, a western state transportation staff member expressed that they far less likely to receive grants for connected vehicle deployments because they are a rural area. In addition, decision makers in New Jersey feel like they have missed their opportunity to participate in connected vehicle deployments since they were not among the first test sites and now need guidance of how to get involved.

4. ***Focus on education and information dissemination.*** Across the board, this study showed the need for education to inform the public sector, trucking industry, and end users about connected vehicle technologies and the benefits they can achieve. The ITS Professional Capacity Building (PCB) programs, American Association of State Highway and Transportation Officials (AASHTO), American Public Transit Association (APTA), the American Trucking Association (ATA), and other organizations were suggested as means to spread the word about the future connected vehicle environment.

5. ***Support resolution of governance issues including security, privacy, and adherence to standards.*** Resolving governance issues appears vitally important to positioning the automotive Original Equipment Manufacturer (OEM) community to move forward. OEMs have practical implementation concerns regarding resolution of security issues and privacy policy as well as how adherence to standards will be enforced and how evolution of standards will be managed in the future. Governance is the mechanism that both allows and insures competitors are able to successfully collaborate - not only among themselves but also with others to bring to fruition the promise of connected vehicle technology.

6. ***Secure the future with supporting commitments from other U.S. Government agencies.*** The GPS system and 5.9 GHz spectrum allocation are foundational elements for connected vehicle technology. Automotive hardware has an unusually long life with limited or no ability to upgrade it. Therefore, the GPS system must be committed to supporting, without degradation, legacy automotive hardware. Similarly, it is imperative that adequate spectrum be allocated for connected vehicle technology. The OEMs are concerned that the failure of either of these would compromise the success rollout of the technology.

7. ***Recognize that competing technologies will temper consumer, trucking industry and OEM enthusiasm for a connected vehicle technology rollout.*** New camera and sensor based technologies that offer some of the benefits of connected vehicle technology are increasingly available as options or even standard equipment. It is important to establish a compelling vision for the public of the unique and longer term benefits of building a connected vehicle environment. The availability of these new technologies in conjunction with the incremental price paid for connected vehicle technology may discourage investment in connected vehicle equipment

Executive Summary

8. ***Reduce the likelihood for long-term risk aversion by establishing incremental successes with connected vehicle pilots and demonstrations.*** Risk for connected vehicle implementation is a concern for all stakeholders, but especially state and local agencies. Shortcomings of connected vehicle implementation could result in pushing innovators and early adopters toward the late majority for technology acquisition.

9. ***Consider ways to provide federal support for continued operations and maintenance of existing and future ITS infrastructure and systems.*** The post-hoc analyses showed that ITS benefits from deployed systems are maintained consistently over time. Expansions do not always offer the same benefits as initial deployment, but economies of scale usually drive down the costs of these expansions. In any case, providing funding to support continued operations and maintenance is generally a wise investment.

10. ***Define national guidelines for connected vehicle implementation*** to minimize incompatible or duplicate systems from being developed and to ensure a consistent deployment approach. The public sector is looking for national leadership in the connected vehicle environment, with deployment guidance as well as in setting standards. The decision making process is more complex, with many more actors in this environment, and states are concerned about unfunded mandates. The public sector is also looking for the business case and is concerned about who would pay for road side infrastructure in a connected vehicle environment. For companies such as long-haul trucking companies that operate across the U.S., the need to purchase several redundant technologies would be a significant burden.

These considerations necessarily focus upon the needs of the three stakeholders interviewed: public sector transportation agencies, automobile manufacturers, and the trucking industry. There is a fourth stakeholder whose views have not been included in this study due its research scope - the end-user of connected vehicle technology. The Connected Vehicle Safety Pilot Program currently underway in Ann Arbor, MI will test performance, evaluate human factors and usability, observe policies and processes, and collect empirical data to present a more accurate, detailed understanding of the potential safety benefits of connected vehicle technologies in a real-world implementation. The data from this pilot will be critical to supporting the 2013 National Highway Traffic Safety Administration (NHTSA) decision on vehicle communications for safety and future decisions on connected vehicle technologies.

1 Introduction

The Intelligent Transportation Systems (ITS) Joint Program Office (JPO) is placing increasing emphasis on transferring ITS technology from research to deployment, and on accelerating the rate of ITS technology adoption. More recently, the JPO expanded and broadened its technology transfer efforts to accelerate deployment of both current and near-future technologies, such as the applications associated with the connected vehicle program. The Longitudinal Study of Implementation provides an important foundation for these JPO efforts.

The objectives of the study were to answer the following research questions:

- What are the most important factors that influence adoption, growth, replacement or cancellation of ITS technology? Do these factors differ according to ITS implementation phase?
- Does expansion of an ITS deployment produce a higher level of benefits? Do benefits trail off as the ITS deployment is operated for a long period?
- What information and delivery methods best support stakeholders in making ITS planning, implementation, O&M, and replacement decisions?
- What can the U.S. DOT do to accelerate ITS technology implementation? What is the applicability of these strategies to connected vehicle advances and next generation ITS?

The Longitudinal Study builds upon a body of existing work related to decision factors influencing ITS adoptions, growth, maintenance or decline within the public and private sectors. The Longitudinal Study goes beyond the current state of knowledge through an interview-based approach to further analyze decision factors; a post-hoc set of studies reviewing deployments, costs, and benefits at early ITS deployment sites; and a workshop and analysis of how to present cost and benefit information in a way that best informs and influences decision-makers.

Building on almost 20 years of ITS deployment experience, Intelligent Transportation Systems (ITS) is at a crossroads, with first generation ITS technologies "at a saturation point" for mature ITS applications, especially in the large metropolitan areas across the United States. As we move forward toward the connected vehicle environment and coordinated operations system envisioned for the future, understanding the motivating factors used by state and local agencies, automobile manufacturers, the commercial vehicle industry for adopting ITS technology is of critical importance.

2 Methodology and Process

The Longitudinal Study of ITS implementation (LSI), conducted by Noblis and its partners, the American Transportation Research Institute (ATRI), Merriweather Advisors, and Cambridge Systematics, provides a foundation that captures the state of knowledge for motivating factors influencing ITS adoption, maintenance, and growth. The study moves beyond the current state of knowledge to understand the motivating factors for adopting technology and for continuing its use and deployment through:
- a comprehensive literature review on technology innovation;
- an interview-based approach to further analyze decision factors;
- post-hoc studies reviewing deployments, costs, and benefits at early ITS deployment sites;
- and a workshop focusing on considerations for future ITS policy and initiatives.

The four key tasks of the LSI are presented in Figure 2-1.

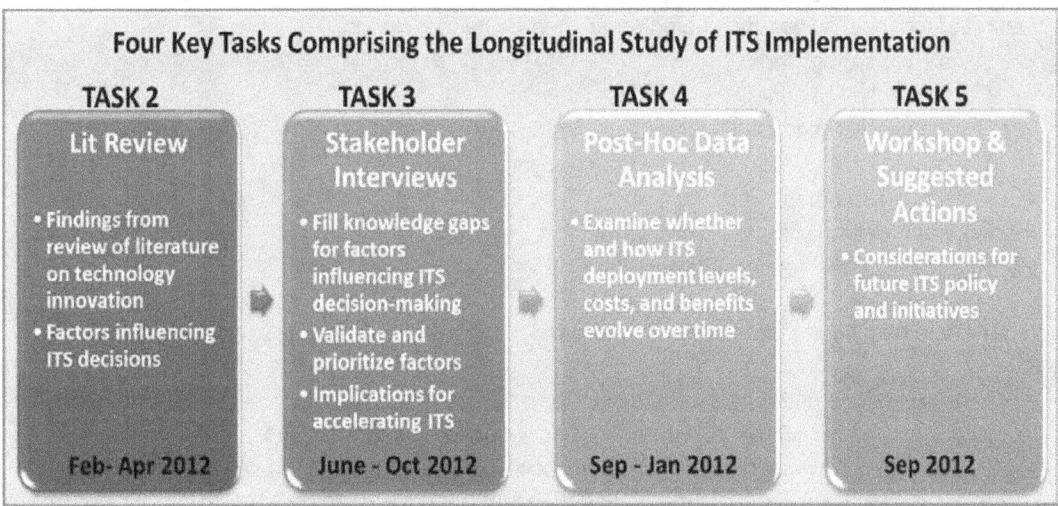

Figure 2-1: Longitudinal Study of ITS Implementation (LSI) Key Tasks *(Source: Noblis 2013)*

2.1 Literature Review

The review of existing literature and the ITS deployment tracking surveys served as the first step in conducting the Longitudinal Study of Implementation. The full document, *Review of Existing Literature and Deployment Tracking Surveys: Decision Factors Influencing ITS Adoption* can be found at http://ntl.bts.gov/lib/45000/45600/45616/FHWA-JPO-12-043_v2_Final_508.pdf. A key goal of this review effort was to identify among existing literature the motivating factors that influence how and why transportation agencies adopt, expand, maintain, cancel, contract (reduce), or deselect technologies. A second goal of the review effort was to highlight the gaps and needs in knowledge for ITS adoption, expansion, maintenance, and decline. The third goal of the review effort was to identify new adopters and others within the expansion, maintenance, or decline phases for ITS. This set of agencies served as a starting point for subsequent tasks within the Longitudinal Study.

In conducting the literature review, Noblis built upon previous JPO-sponsored work performed by a number of organizations. The review considered and offered special attention to materials that expressed technology deployment to support transit and truck operations, public safety, and maintenance and construction operations, especially in areas that extend beyond the market areas covered in the background material. The study team also explored models of technology adoption, and examined other work on the theory of innovation, including international efforts. These efforts were important to establishing the proper framework for the subsequent interviews.

2.1.1 Webinar Results

The webinar served as the first phase in addressing a key knowledge gap identified during the Literature Review, the lack of research on factors influencing subsequent decisions to grow, maintain, contract or cancel specific ITS technologies or systems. Whereas for the adoption decision, participants were polled on each set of factors, the time constraints of the webinar supported only a single question for the three other types of ITS decisions. Consequently, the previous four sets of factors were downsized to seven factors and an 'other' category. These seven were selected based on what researchers anticipated as being most important and include:

- Price – technology factor
- Demonstration of benefits – technology factor
- Standards/interoperability –technology factor
- Users' acceptance/attitudes –user/market factor
- Political environment –external factor
- Budget/Funding Sources –external factor
- Agency Priorities –external factor

Findings of note concerning the decisions to **expand, maintain, contract or cancel** ITS projects include:

- The greatest number of participants selected "price of the technology" as the most important factor in the decision to expand, maintain, contract/cancel ITS projects.
- For decisions to expand systems, "users' acceptance/attitudes" was cited as most important by more than three times as many responses than "budget/funding sources." However, when maintaining or contracting/cancelling ITS, "budget/funding sources" is cited as often as "users' acceptance/attitudes."
- Whereas the political environment was not a major factor for ITS system expansion, a few participants did cite this as the most important factor in decisions to contract or cancel ITS technologies.
- Finally, "demonstration of benefits" is a far more prominent factor when deciding to contract/cancel a system compared to decisions to adopt, expand, or maintain ITS systems or technologies.

The webinar participants were also asked if they were interested in future participation in the study with a brief phone interview. These phone interviews were used as preliminary screening interviews to select locations for in-depth interviews and site visits.

2.2 Stakeholder Interviews

To make efficient use of resources and make each visit effective, the Noblis team selected regions/locations that include multiple deployments of various types. Sites were selected to provide a representative sample across as many different potential decision factors and influences as possible. These criteria included:

- Different ITS applications supporting transit and truck operations, public safety, and maintenance and construction operations. Electronic toll payment, highway data collection, vehicle data collection, traffic control software, and signal priority and pre-emption were also of interest.
- Technological maturity of the application at the time of decision (innovation, proven, mature, legacy).
- Geographic distribution across the country.
- Problems and conditions factors such as congestion, air quality attainment, local funding and resources
- Type of ITS implementer, from a pioneering region with a history of early deployment of innovations, to mainstream imitators that implement proven technologies, to laggards with only recent implementations.
- Other external factors and differentiators, including size and type of agency or region, institutional organization and regional cooperation, economic growth or contraction, and type of funding used (earmarks, formula grants, ITS initiative or early deployment grant, etc.)

Key to the Noblis team approach to address future as well as past decisions was to capture new/emerging technology as well as traditional ITS technologies. The team included one small area or rural site to capture differences in deployment factors in places other than the large urban areas captured by ITS Deployment Surveys.

Figure 2-2 provides an overview of activities implemented to complete the stakeholder interview efforts. The green describe the activities, while the blue spheres highlight the types of organizations represented through surveying activities. As illustrated in Figure 2-2, the public sector surveying was a far more intensive process including multiple traditional modal transportation entities as well as law enforcement, Emergency Management Systems (EMS), and toll authorities. A four-step process was applied for selecting interviews, and the interview process for this sector included site visits with individual interviews and panel sessions. Trucking and auto manufacturer interviews were more streamlined and consisted of interviews with key decision makers within companies.

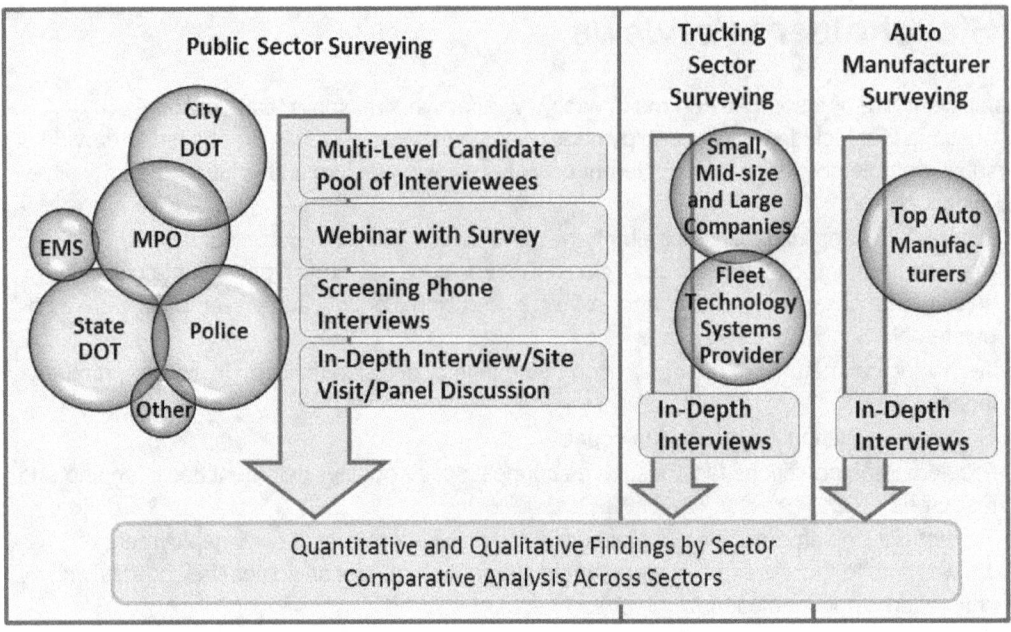

Figure 2-2: Activities Implemented and Organization Types Interviewed to Address Concepts, Factors, and Issues Related to ITS Decision-Making *(Source: Noblis 2013)*

The Noblis team conducted in-depth telephone interviews with four distinct trucking industry stakeholders. These interviews were designed to gather information from decision makers involved in the adoption and deployment of various in-vehicle and trucking systems technologies. Interviewees were selected to ensure a diversity of trucking perspectives based on company size and carrier type as well as industry perspective specific to technology decisions.

The Noblis team also conducted interviews with key representatives within three major automobile manufacturers. The goal of these interviews were to explore implementation of ITS by the auto manufacturing industry focusing on issues of relevance to the Original Equipment Manufacturers (OEM) industry in moving toward connected vehicle deployment. Individuals interviewed represented leadership at each organization with responsibilities in safety, engineering, and research. Eight OEM individuals participated among the three corporate meetings, each lasting approximately an hour.

Upon the completion of the webinar and interviews, a cross-cutting analysis of the responses was conducted. Qualitative as well as quantitative results were analyzed to gain insights into the public sector responses. Overall themes and similar findings across all of the public sector interviews were distilled from the individual responses. Responses were also compared and contrasted between different agency characteristics and ITS application areas. Key patterns and themes were culled from interviews, and these were used to generate insight and fill the knowledge gaps identified in the Literature review. The instruments used for this process included a preliminary e-survey, a screening phone interview survey, and an in depth, generally in-person interview, all found in Appendix C Interview Instruments.

2.3 Post-Hoc Analysis

The post-hoc analysis was conducted to assess the longer-term impacts and benefits of investment in ITS. The goal of this effort was to examine how the performance of various systems have changed over time either due to expansion/enhancement of the systems, or changing traffic patterns or traveler behavior. When possible, this effort identified those factors that have contributed to or hindered the success and expansion of deployments. Evaluations include assessment of:

- a transit traveler information system in Portland, Oregon
- a ramp metering deployment in Kansas City,
- high occupancy toll (HOT) lanes in Minneapolis/St. Paul, and
- an arterial management system in Phoenix, Arizona.

Figure 2-3: ITS Technologies evaluated as part of the post-hoc analysis *(Source from left to right: Think Stock 2013; Wikimedia Commons/Patriarca12 2013 ;Minnesota DOT 2013; Think Stock/David De Lossy 2013)*

Each of these sites was the subject of a past evaluation effort intended to assess the impacts, benefits, and/or costs, often closely following the original deployment of the selected strategies. The original results of these studies were used as the baseline for conducting this longitudinal evaluation. Various data representing both traffic system performance (e.g., speeds, volumes, travel times, transit ridership) were compiled from reliable archived data sources for the periods between the previous evaluation effort and the current day. These were combined with data covering the deployed systems, including enhancements in technology or expansions in coverage, as well as changes in system costs.

2.4 Public Workshop

On September 25-27, 2012, the Intelligent Transportation Systems (ITS) Joint Program Office (JPO), U.S. Department of Transportation (U.S.DOT) hosted a free public meeting and webinar to provide updates and promote discussion on the Connected Vehicle Safety, Vehicle-to-Infrastructure, and Testing programs, with a special session on Day 3 titled, "Applying Past Experience to Achieve Future Success." Section 7 summarizes the results of the Day 3 session which was devoted to discussion of findings of this study. The goal of the discussion was to identify what we have learned from past ITS deployments that can help achieve success for the future connected vehicle environment.

3 Public Sector Perspective

The Noblis team conducted five public sector site visits and two additional site reviews as part of this study. The purpose of these site visits was to explore in detail with decision makers the factors influencing ITS decision made by the public sector, and their perspectives on a number of issues facing their agencies. The interviews solicited demographic information, attitudes towards ITS technology adoption, the environment of a recent ITS decisions, and quantitative ratings of importance for the four sets of factors influencing ITS decision-making. Up to five open ended issues were also discussed with each interviewee to better understand their perspectives on current ITS decision making and their perceptions of the next generation of ITS deployments. Discussions were based on time availability and the interests of the interviewee. These free-form discussions issues were also replicated during a panel session conducted at each site visit.

Figure 3-1: Site visit to Boise, Idaho *(Source: Noblis 2012)*

The team conducted multi-day site visits in Boise, Phoenix, Tucson, Atlanta, and Baltimore/Washington with state, regional, and local transportation agencies. In some regions, police, fire, safety, and EMS decision makers were interviewed as well. Each site visit involved six to eleven individual interviews, followed by a panel discussion. The purpose of the panel session was to bring together the many constituencies involved in decision-making to discuss varied perspectives. In addition, the Seattle region in Washington State and the New Jersey Department of Transportation were interviewed in person and over the phone to bring the total to six states with in-depth interviews in seven regions. These in-depth stakeholder interviews are geographically illustrated in Figure 3-2.

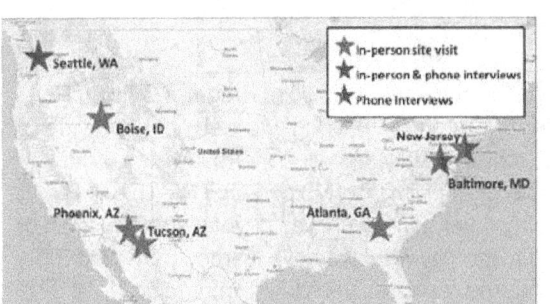

Figure 3-2: Public Sector Visit Locations *(Source: Noblis 2013)*

Including the panel sessions, there were 46 total participants, representing state and local transportation DOTs, transit agencies, law enforcement, emergency services, and planning regional organizations.

Appendix C contains the individual site visit summaries as well as the interview and survey instruments used in the study. The presentation contained in this section blends and summarizes the findings gathered during the in-depth interviews across sites, including the quantitative survey data on key decision factors and barriers. Information from the individual sites is used where appropriate to provide examples and illustrate key points.

3.1 Importance of ITS Decision Factors

Interviewees were asked to rate the importance of each of the 23 factors surfaced during the literature review by stage of ITS decision-making. The set of 23 factors were organized among the four categories of:
- technology and application factors,
- implementing organization factors,
- user and market factors, and
- external factors.

Figure 3-3 depicts the 23 factors organized by category. The three stages of ITS decision making were defined as shown in Figure 3-4.

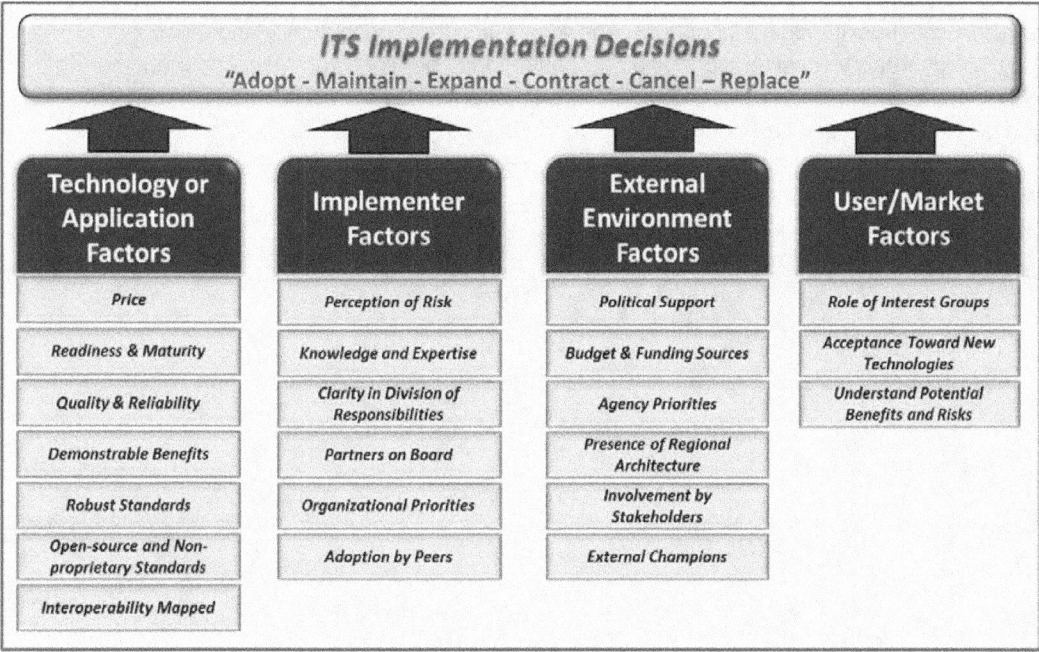

Figure 3-3: Sets of Factors Influencing ITS Implementation Decisions *(Source: Noblis 2013)*

3 Public Sector Perspective

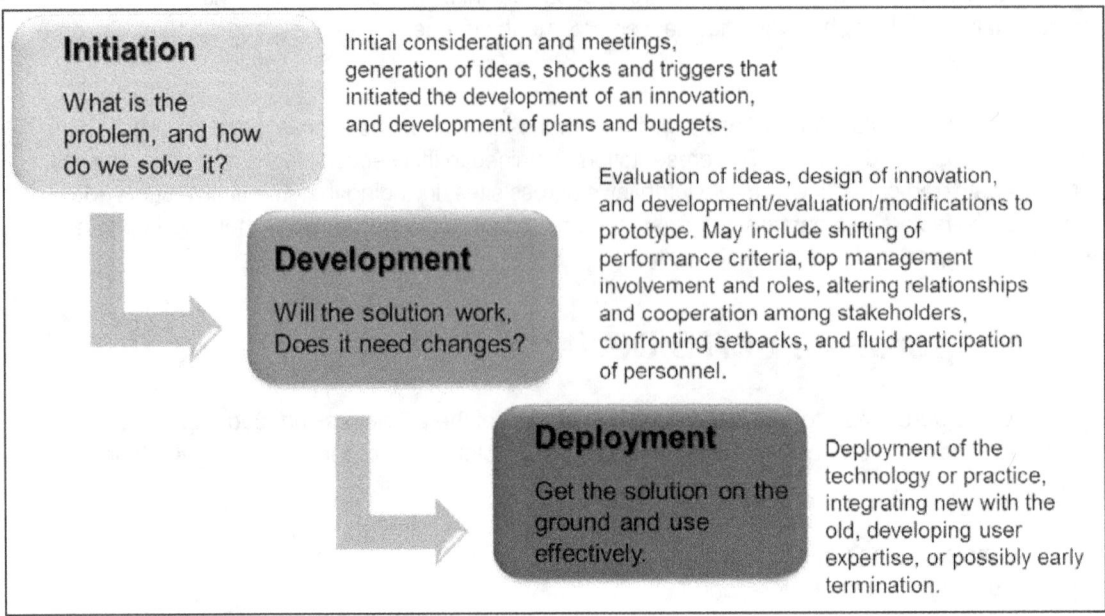

Figure 3-4: Three Stages of ITS Decision Making *(Source: Noblis 2013)*

Table 3-1 provides detailed rating statistics by specific factors. Across the three phases of ITS implementation, the most important technology and application factor was quality and reliability, followed by interoperability considerations and demonstration of benefits. The most important among implementing organization factors are having partners on board and organizational priorities. The most important external factor was clearly budget and funding sources. This factor was also the highest rated among all factors on average across the three phases of ITS implementation. These factors were on average important across all regions interviewed.

3 Public Sector Perspective

Table 3-1: Detailed Importance Rating Statistics for Factors Influencing the Decision to Implement ITS *(Source: Noblis 2013)*

Factors	Sub-category	# Interviewees: 30	Statistics Computed — Average | Range (minimum - maximum)		
		\multicolumn{4}{c}{Importance of Factors/Subcategories (1-10)}		
		Initiation	Development	Deployment	Avg
Technology and Application	price	5.8 | 1 - 10	7.0 | 4 - 10	7.4 | 1 - 10	6.8
	readiness and maturity	6.4 | 2 - 10	7.3 | 3 - 10	7.6 | 1 - 10	7.1
	quality and reliability	6.7 | 1 - 10	7.7 | 1 - 10	8.9 | 7 - 10	7.8
	demonstration of benefits	7.2 | 2 - 10	7.3 | 5 - 10	7.7 | 1 - 10	7.4
	sufficiently robust standards	5.7 | 2 - 10	6.8 | 4 - 10	6.7 | 1 - 10	6.4
	open source software	6.1 | 1 - 10	6.7 | 1 - 10	6.2 | 1 - 10	6.3
	interoperability mapped	6.5 | 1 - 10	7.9 | 5 - 10	7.8 | 1 - 10	7.4
Implementing Organization	perception of risk	6.2 | 1 - 10	6.6 | 1 - 10	6.8 | 1 - 10	6.5
	knowledge and extertise available	6.4 | 3 - 10	7.6 | 2 - 10	8.0 | 2 - 10	7.3
	clarity in division of responsibilities	6.0 | 1 - 10	7.0 | 2 - 10	8.0 | 2 - 10	7.0
	partners on board	7.0 | 1 - 10	7.6 | 3 - 10	8.0 | 5 - 10	7.5
	organizational priorities	7.1 | 3 - 10	7.3 | 3 - 10	7.7 | 1 - 10	7.4
	adoption by peers	5.1 | 1 - 9	5.7 | 1 - 10	5.9 | 1 - 9	5.6
User and Market	the role of interest groups	5.7 | 1 - 10	5.3 | 1 - 10	5.4 | 1 - 10	5.5
	end-user acceptance	6.6 | 1 - 10	7.1 | 1 - 10	8.3 | 1 - 10	7.3
	end-user awareness and understanding of benefits	6.5 | 1 - 10	7.1 | 1 - 10	8.4 | 1 - 10	7.4
External	local, regional, state, federal political support	6.8 | 2 - 10	7.1 | 1 - 10	7.0 | 3 - 10	7.0
	budget/funding sources	7.8 | 2 - 10	8.4 | 1 - 10	8.3 | 1 - 10	8.2
	overarching agency priorities	7.1 | 3 - 10	7.5 | 3 - 10	7.7 | 1 - 10	7.4
	presence of a regional architecture	5.7 | 1 - 10	5.9 | 1 - 10	5.9 | 1 - 10	5.8
	involvement in the project by stakeholders	7.2 | 1 - 10	7.7 | 3 - 10	7.8 | 1 - 10	7.5
	external champions	5.8 | 1 - 10	6.0 | 1 - 10	6.4 | 1 - 10	6.1
	links with universities and research centers	5.1 | 1 - 10	5.0 | 1 - 10	4.3 | 1 - 9	4.8

States surveyed: Idaho, Arizona, Maryland/DC, Georgia, New Jersey, Washington State

Importance of Factors by Phase of ITS Implementation

One of the gaps identified by the literature review was that motiving factors may change according to the phase of ITS implementation. To address this gap, the study asked public sector stakeholders to rank the decision factors by phase of implementation: initiation, development or deployment.

During the initiation phase, budget and funding was most critical (rating of 7.8). Demonstration of benefits and involvement in the project by stakeholders comes in second, both with an average rating of 7.2 (Figure 3-5).

3 Public Sector Perspective

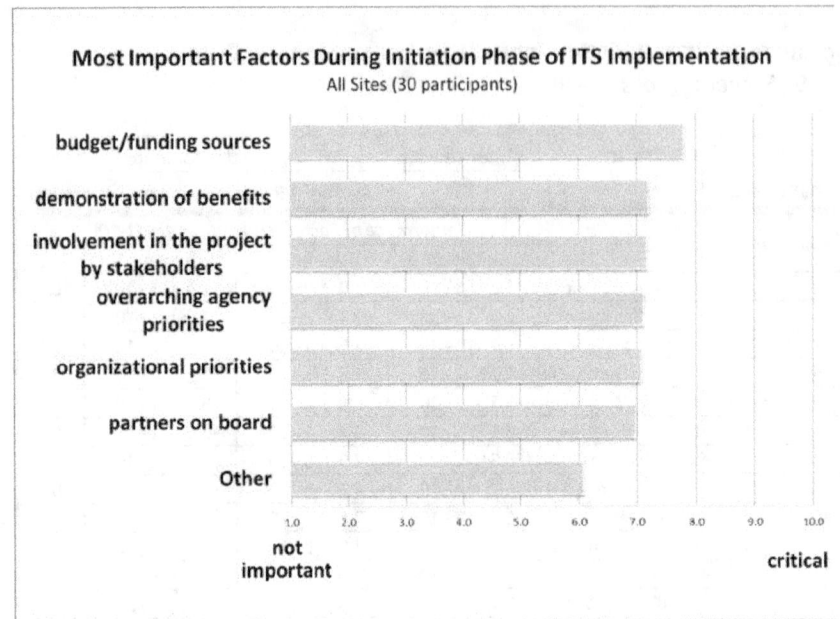

Figure 3-5: Most Important Factors during Initiation Phase of ITS Implementation *(Source: Noblis 2013)*

Moving to the development phase, budget and funding continued to be important with an even higher rating, that of 8.4 out of 10. During the development phase, interoperability came in second, with an average rating of 7.9 (Figure 3-6)

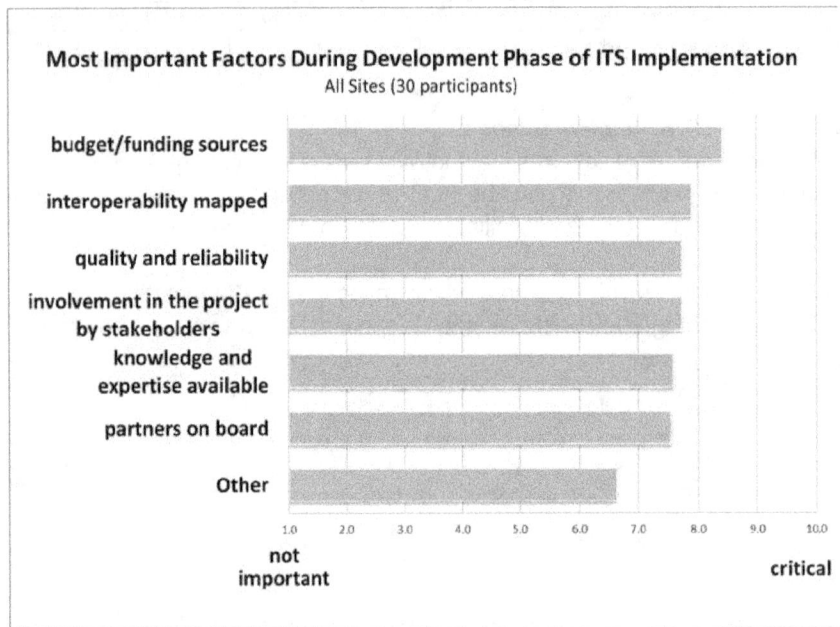

Figure 3-6: Most important Factors during Development Phase of ITS Implementation *(Source: Noblis 2013)*

In the deployment phase, quality and reliability (8.9 rating) and end-user awareness/understanding (8.4 rating) superseded budget and funding (8.3 rating). During this phase of ITS implementation, a number of organizational factors took precedence including staff knowledge and expertise, clarity in division of responsibilities, and having partners on board (all with 8.0 rating). These results are shown in Figure 3-7.

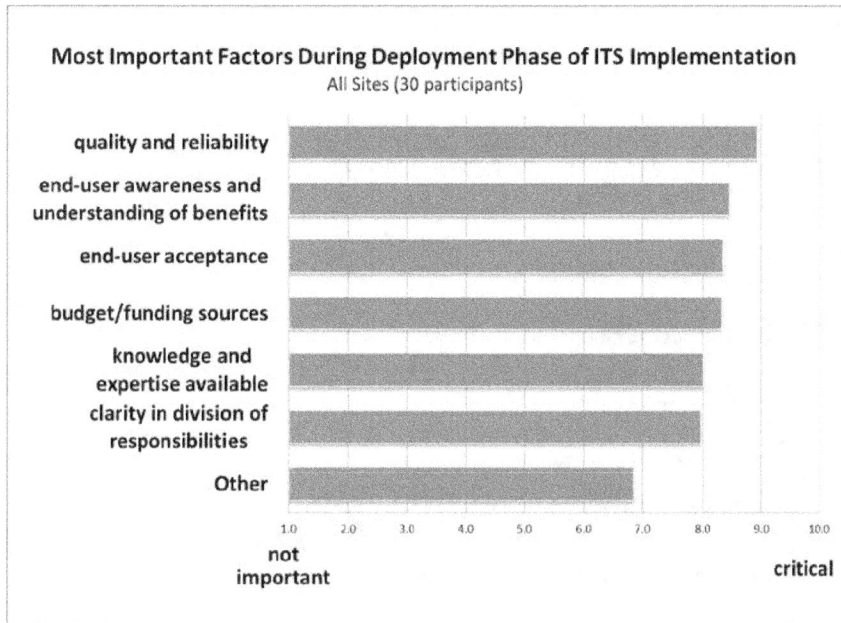

Figure 3-7: Most Important Factors during Deployment Phase of ITS Implementation *(Source: Noblis 2013)*

These findings were confirmed by qualitative analysis of the interviews. For adoption and growth decisions that discussed technology and application factors, the largest percentage cited demonstration of benefits (55%) as a factor in the decision to adopt or grow the technology. This percentage was smaller for other types of decisions, such as replace (27%), cancel (20%) and not select a technology (6%). Across the board, for these other types of decisions, quality and reliability of technology was cited most often as the most important technology factor.

Least important across all categories of factors were links with universities and research centers (4.8 rating) followed by the role of interest groups (5.5) and then adoption by peers (5.6). Two of these findings differ from the qualitative assessments. For example, in Idaho, adoption by peers was a significant topic of interest. Also, for New Jersey, the links with universities and research centers was a critical growth area that is expected to support needed workforce development. These factors are of importance to some regions but clearly not all.

3.2 ITS Implementation Barriers

Participants were also asked to rate the importance of specific barriers to implementation of ITS technologies or systems. This assessment was focused specifically on identifying the factor that was

most instrumental in a decision not to implement an ITS technology. The barriers were organized along six categories with three factors in each category:
- available information (knowledge) barriers,
- technical barriers,
- legal and regulatory barriers,
- financial and economic barriers,
- cultural and societal barriers, and
- decision-making barriers.

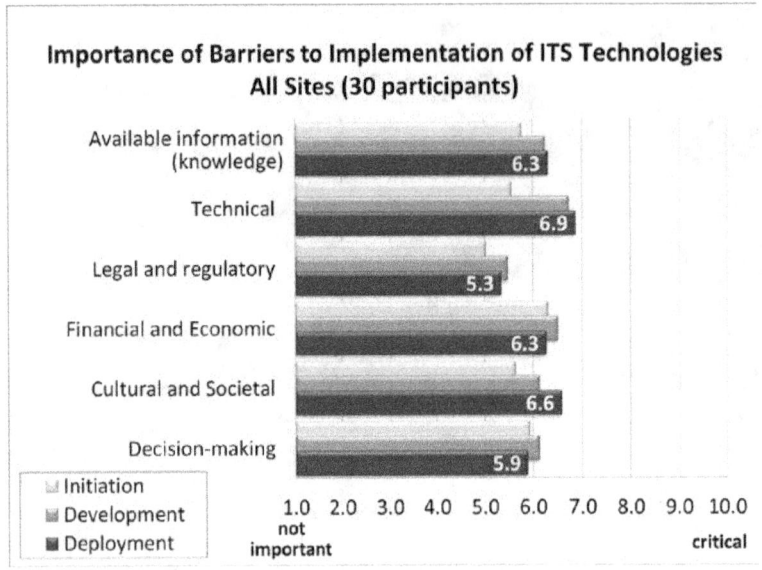

Figure 3-8 Rating of Importance of Barriers to ITS Implementation *(Source: Noblis 2013)*

Figure 3-8 summarizes the average outcomes of rating across the specific barriers in each of the six factor sets. Average ratings are all above a value of 5.0 out of ten, implying that every factor set listed is important to some constituency. The results show that as an ITS implementation moves from initiation, through development, and into deployment; knowledge, technical, and societal barriers become more critical. Legal and regulatory, financial and economic, and decision making barriers are more important during the development stage compared to the initiation or deployment stages of ITS implementation.

Financial and economic factors were cited as the most critical set of barriers during the initiation phase followed by decision-making and information availability challenges. During development and deployment, technical factors become the most critical barrier to ITS implementation.

When the barriers were broken out by specific factors, costs and funding availability were ranked as the most critical barriers to ITS initiation, followed by lack of interoperability and lack of qualified personnel. In fact, during the development phase, lack of funds and lack of interoperability were rated as equally critical barriers. During the deployment phase of an ITS implementation, the most critical barrier was interoperability followed by the inability to devote staff to the activity or lack of qualified personnel.

The rating responses varied significantly from participant to participant. For the majority of factors at least on interviewee rated the factor as not important (value of 1), while at least one individual rated the same factor as critical (value of 10).

The surveying also confirmed that there are significant differences regionally and across individuals within a region as far as the importance of specific factors. For example, New Jersey interviewees cited workforce knowledge and expertise as extremely critical while many in Georgia rated this far lower.

3.3 Support for ITS Investment Decisions

The site visit interviews enabled interviewees to expand on five topics related to their ITS decision making process, as described within this section. These topics were:
- Evolution of the ITS decision making process
- Role of Performance Measurement in ITS
- Prioritization in a constrained economic environment
- Interoperability and Integration
- Awareness and implications of connected vehicle technology

3.3.1 Evolution of the ITS decision making process

ITS is inherently an operational technology, so there has been some difficulty determining where it fits in to the overall transportation planning process. It is clear from the public sector that the structure of the ITS decision-making process has evolved over the last decade to become much more mainstreamed and streamlined into the transportation planning process. However it was also reported that the overall complexity of the process has grown in several ways.

In the past, the decision to implement ITS system came from a desire to try out a new technology, often due to an event. For example, in the early 1990s, the City of Atlanta deployed new ITS technologies as part of the preparation for the 1996 Olympic games. Now, ITS technologies are integrated into traditional transportation projects and many state and local jurisdictions have dedicated staff and funding for ITS implementations.

With the evolution of ITS into mainstream transportation decisions, overall complexity has grown in several ways. There are now many more agencies and stakeholders making ITS decisions, including participation from law enforcement, fire and EMS. In addition, regional interaction is an important consideration in ITS deployment decisions. New implementations follow the regional architecture and ensure that all regional stakeholders are on board. To ensure regional interoperability of ITS systems, stakeholders in Idaho, Washington, Montana, Wyoming and several other states have implemented the Northwest Passage Pooled Fund Study to ensure consistency in their transportation implementations and operations.

Figure 3-9: Traffic Management Center in Boise, ID *(Source: Noblis 2012)*

Another major change in the ITS decision making process is that peer influence is more significant now than with the first wave of early ITS adopters. ITS innovation is happening in many areas, but documented benefits specific to regional characteristics can be lacking. Therefore, many agencies look to their regional peers for the 'right fit' in implementing ITS. Multi-state, multi-modal coordination plays a significant role in accelerating impact of ITS deployment as in the case with 511 deployments. As peer coordination becomes increasingly more significant, is important to note that the designation of who serves as a peer differs depending on the region and reflects size and congestion levels.

After a decade of focusing on ITS deployments and a period of constrained economic funding, there has been a shift from deployment of ITS systems to the maintenance and operations of these systems. In addition, the lack of in-house expertise and rapidly changing technologies pose a barrier to adoption of new ITS technologies as ITS staff must focus on the operation of their existing systems. With a greater attention on operations and maintenance, there is a clear need for education and training for ITS staff on operations of ITS.

> "Shift from deployment to operations requires more training"

In reviewing ITS deployment stories, many agencies described how they alternate between being innovators and early adopters for some technologies while being the late majority for other technologies. A law enforcement agency recounted how the disappointments with one technology for which they were innovators soured upper management for more than a decade on other ITS innovations. The pendulum is now swinging back toward the early majority for ITS adoption.

3.3.2 Role of Performance Measurement in ITS

With the new legislative provisions for performance reporting through MAP-21, states will be given funding based upon performance. State agencies are apprehensive of what this will mean for their ITS programs in terms of obtaining funding and what performance measures will be required. State agency staff expressed that they are looking for Federal guidance for better, clearer information on what performance metrics can most effectively be measured to manage by performance.

Figure 3-10: Maryland DOT Traffic Management Center *(Source: Noblis 2012)*

State and local stakeholders recognize that performance monitoring can be very effective for proving benefits and have the potential to bridge the divide between technical and non-technical decision makers. However, these agencies voiced concern that performance measures can be costly and sometimes difficult to measure. The lack of staff and technical expertise is growing concern of state and local agencies. With the current constrained economic environment, it is difficult to find the budget to hire expert technical staff that can perform data analysis.

One agency in Washington State noted that it is costly to conduct performance monitoring even with a good system already in place. They have plenty of data being collected, but no time or expertise for data analysis. In addition, many stakeholders noted that it is difficult to measure the performance of some essential ITS technologies such as dynamic message signs, cameras, and training for personnel.

"Data rich, analysis poor"

In Arizona, several stakeholders noted that ITS technologies implemented for a particular purpose may also aid in a different way. For example, while traffic agencies implement cameras primarily for traffic monitoring, law enforcement and rescue officials with access to the camera views can also benefit from seeing the video feed from an incident. Therefore, ITS technologies have the potential to be a "force multiplier" for benefits and these should be captured in some way through performance measures.

3.3.3 Prioritization in a constrained economic environment

ITS has the reputation of being low cost, high benefit option but continues to face tough competition. On one hand, many decision makers understand the value of ITS solutions in a constrained economic environment. On the other hand, many state agencies expressed the challenge that politicians seek to deliver high profile construction projects and ribbon cuttings, which is rarely the case with ITS projects. The investment in ITS also represents an ongoing investment in operations and maintenance which is often at the mercy of the software developers.

Several agencies expressed that federal requirements are making it difficult for local jurisdictions to apply for federal funding. Local jurisdictions have a difficult time applying for federal funding because of the requirements associated with the funding.

Figure 3-11: Atlanta, Georgia HOT lanes (Source: Noblis 2012)

State agencies have also expressed issues involving using federal funding because of the 'strings' that come attached. For example, NJDOT returned a large sum of money after they were unable to use the funding for their project. They received $13 million for the construction of a new adaptive signal control system but were not permitted to use any of the funds to do the design of the system. NJDOT could not obtain design funds and therefore was not able to use the construction funds.

"Federal funding has too many strings attached"

In a constrained economy, sharing resources has been important to continuing ITS growth. For example, traveler information from a transportation agency can be shared with law enforcement, fire, and emergency response. One successful deployment is Tucson's fiberoptic backbone project where several agencies, including police and fire, can use the DOT's communications infrastructure. Another example is Idaho's emergency call center that is used by Idaho's Transportation Department, Idaho State Police and Idaho Health and Welfare.

3.3.4 Interoperability and Integration

Compatibility and interoperability become more critical as the installed base increases. Looking back at the last 15 years of ITS, deployers are very concerned about existing systems working together. As new systems are deployed, organizations must consider whether these systems will be compatible with their existing "legacy" systems. Interoperability is now a consideration locally, regionally, statewide, and between states.

"Think regionally not locally"

While multimodal integration is of high priority to state and local agencies, a desire for interoperable systems has created new cross jurisdictional and institutional issues that need to be addressed. Integration and interoperability of different systems is very important within each agency, between agencies, and with neighboring states. However, IT departments have been reluctant to open up internal networks to other agencies due to security concerns. In addition, funding limitations may lead to agencies deploying systems that are not interoperable.

The presence of a strong regional organization that brings together stakeholder agencies is tremendously effective in promoting a coordinated adoption, maintenance and growth of ITS. In Arizona, AzTech is a regional organization that helps promote ITS solutions throughout the state. Similarly, the Northwest Passage Pooled Fund Study brings together organizations in ten states to promote ITS and ensure regional interoperability.

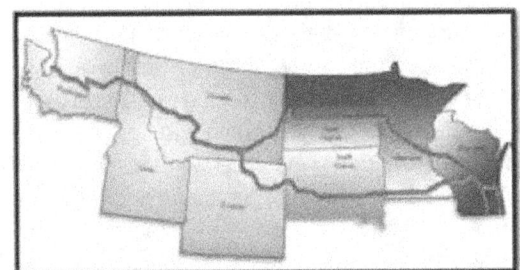

Figure 3-12: Northwest Passage *(Courtesy of Northwest Passage Pooled Fund Study 2013)*

3.3.5 Awareness and implications of connected vehicle technology

Many organizations reported that they are unclear about the connected vehicle program. Stakeholders are unsure of what the proposed applications are, particularly in the vehicle-to-infrastructure side, and what the implementation period for these applications would be.

"How do I get involved? I feel like I missed the boat"

"What is my role?"

Some state and local stakeholders are very involved in the connected vehicle program, such as the Arizona DOT ITS personnel who run a connected vehicle test-bed located in Phoenix. However, other agencies reported that they do not know how to become involved in the program and feel like they may have "missed the boat." These agencies are looking for opportunities and federal guidance for how to become involved in the program.

Decision makers are also unsure of the public sector's role will be in the connected vehicle research program. Agencies can see the value of the program and are very interested in connected vehicle applications but are more aware of the vehicle to vehicle (V2V) applications that can be largely handled by the auto manufacturers. Several of the decision makers reported that they are not very aware of the vehicle to infrastructure (V2I) applications.

Decision makers also want more information on the business cases to support justification. There is a concern that connected vehicle applications would be mandated without federal funding to help implement it. Stakeholders expressed that the federal government needs to provide more demonstrations, training and direction to local and state DOTs.

Figure 3-13: Connected vehicle environment *(Source: U.S. DOT 2013)*

4 Trucking Industry Perspective

The trucking industry interviews led by ATRI investigated the decision making process for technology adoption in the industry. These interviews were designed to gather information from decision makers involved in the adoption and deployment of various in-vehicle and trucking systems technologies. Interviewees were selected to ensure a diversity of trucking perspectives based on company size and carrier type as well as industry perspective specific to technology decisions.

Interviews were conducted with three motor carriers and a nationally-recognized trucking technology provider. The motor carriers included small (5-100 vehicles), medium (100-500 vehicles) and large (500+) fleets that operated predominantly in the U.S. The trucking technology vendor serves motor carriers with tens of vehicles to thousands of vehicles. Interviewees were high-level executives, including two corporate presidents, a chief financial officer/vice president, and a vice-president of safety. This interview confirmed views and practices shared by the motor carriers.

The three motor carrier interviewees shared their recent major technology adoption decisions which included:
- an upgrade of a fleet's onboard telecommunications systems,
- field testing of electronic onboard recorders (EOBRs) and in-cab camera systems, and
- distribution of android-based smartphones to drivers in conjunction with EOBR installation.

Although each technology decision was made to address a number of different organizational objectives, improving operational productivity and efficiency was a uniformly cited reason for deciding to upgrade or install a particular system. Other specific goals mentioned by motor carriers included safety and financial risk management. The technology provider also indicated that many choose to adopt technologies because of compliance and back-office/administrative efficiency.

One motor carrier was adopting a technology, a second was adopting and testing a technology, and the third was upgrading an existing technology. The number of units that were upgraded or installed varied among fleets, ranging from 25 to 200 units. All required an investment ranging from $30 to $100 per unit, while one also involved significantly more in monthly service fees. The technology provider noted that "new installations are typically on a smaller percentage of the fleet compared to upgrades, and both are generally not full-fleet because of cost constraints and fleet service schedules."

Technology related decision-making proved to be more centralized involving a limited number of key personnel for smaller fleets. As fleet size grows, the pool of decision makers expands considerably. The small fleet interviewee reported that the president, chief executive officer (CEO) or vice president of operations propose and research a technology and all three jointly make the decision. The medium-size fleet carrier decision-making involved a few more individuals, while the large fleet carrier decision-making process can "reach from the president to the drivers, with the main process being driven by the executive-level staff." The technology provider confirmed this relationship noting that only a few individuals are involved for smaller fleets, while as many as 20 individuals may be involved in technology acquisitions among large fleet operators. The figure below shows the number of decision makers that are involved based on the size of the trucking company.

4 Trucking Industry Perspective

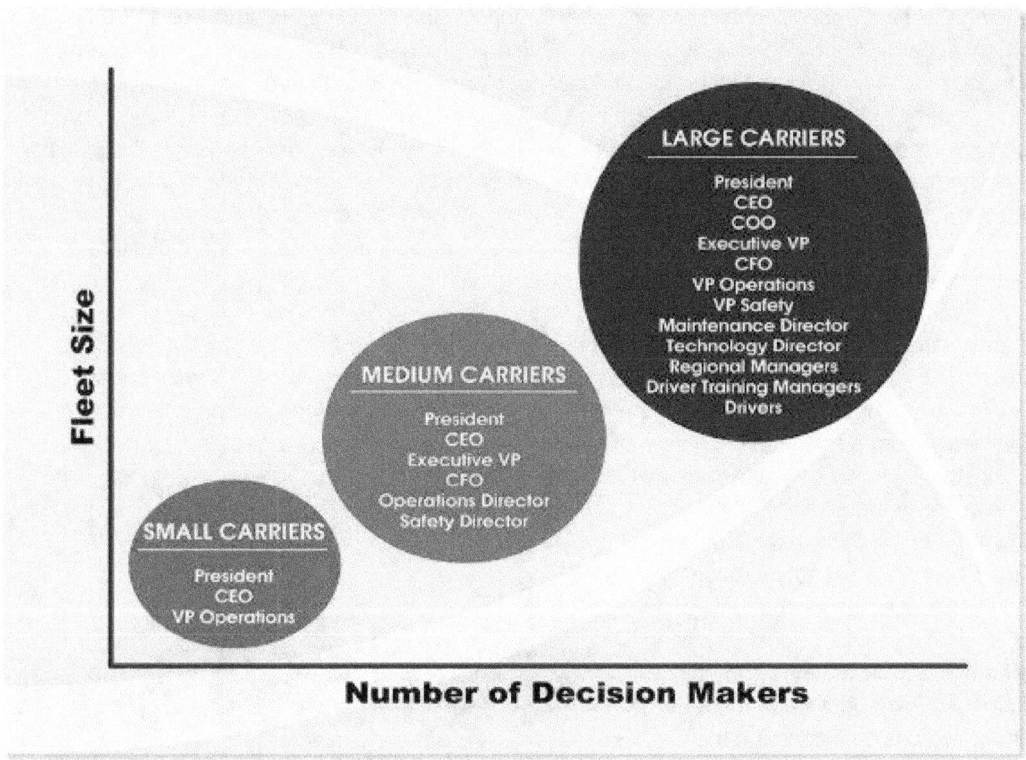

Figure 4-1: Number of decision makers based on fleet size *(Source: ATRI 2013)*

The types of analyses and other resources that carriers employ when selecting a new technology also vary widely between small and large fleets. Large carriers tend to be early adopters of new technologies and are more willing to test a new, possibly unproven system. These large fleets may issue a request for proposals (RFP) from vendors and will complete the necessary analyses in-house. Alternately, small carriers often rely on the vendors to perform the return-on-investment (ROI) and payback period analyses.

One carrier noted that they gather information from other fleets on their experience with a particular device (i.e. "peer to peer" information exchanges). This allows the carrier to learn not only about the benefits of the system, but also any difficulties that were encountered during a real-life deployment.

The interviewees were asked a series of questions regarding the factors that influence their decisions to adopt technology and their company's attitudes toward new technology.

Overall, the most important factors for adopting a new technology were:
- Price of the technology/ Return-on-Investment (particularly for small carriers)
- Compatibility with the existing system (a big concern for large carriers)
- Readiness and maturity of the technology
- Quality and reliability
- Service and support
- System integration and flexibility

Participants were asked to provide examples of things that the public sector currently does well, as well as things that the public sector could do better to support the trucking industry. For all fleet sizes, government sponsorship of technology research, development, and testing is critical. Small carriers rely on financial support (tax credits, grants and low-cost loans) more than large fleets.

Two suggestions on ways that the public sector could increase technology deployment in the industry include mandating certain systems (such as EOBRs) in order to "level the playing field" and to improve truck navigation systems for hazardous materials routing. Another recommendation was to provide funding to roadside enforcement agencies to develop a system to transfer data from trucks to the enforcement personnel.

Participants were also asked about their awareness and attitude towards the connected vehicle program. Industry awareness of the connected vehicle program was very limited. Two of the three carriers interviewed had no knowledge of the initiative, while the telecommunications provider had heard of the program but did not know any specifics. The large fleet was aware of the connected vehicle initiative, but only because they are actively involved in the pilot test of the program. One of the participants noted that industry interest will likely not increase since many existing technologies and tools can provide similar types of services. Below are the three major considerations from the trucking industry perspective for the federal government to consider regarding the connected vehicle environment:

Figure 4-2: Connected freight vehicle *(Source: U.S. DOT 2013)*

- ***Identify the potential economic impacts using real-word cost-benefit, ROI and payback period analyses.*** The private sector needs current and accurate cost information in order to assess the benefits of this initiative. Since the connected vehicle program relies on technologies in both the infrastructure and vehicle, private sector engagement will be crucial to its success.
- ***Address the privacy concerns associated with connected vehicle technologies and data collection/transfer efforts.*** Privacy concerns will be different for both the public and private sectors as well as for stakeholders within each group. Businesses such as the trucking industry will likely have the highest degree of concern since there is the potential for sensitive/proprietary information to be transmitted.
- ***Define a national blueprint to minimize incompatible or duplicate systems from being developed and to ensure consistent implementation.*** Businesses currently operate under a patchwork of regulations that vary state-by-state and stakeholders would find more benefit from a uniform program. Furthermore, for companies such as long-haul trucking companies that operate across the U.S., the need to purchase several redundant technologies would be a significant burden.

The trucking industry publication "Transport Topics" was the most widely suggested media outlet for the public sector to communicate about the connected vehicle program with the trucking industry. Additional communications channels include Intelligent Transportation Systems (ITS) groups, other industry publications such as the Commercial Carrier Journal, and through social media sites.

5 Automotive Manufacturer Perspective

The Noblis team conducted interviews with key representatives within three major automobile manufacturers, all of which are actively involved in the connected vehicle research program. The goal of these interviews was to explore implementation of ITS and related technologies by the auto manufacturing industry focusing on issues of relevance to connected vehicle deployment. The findings from these interviews are organized and presented by topic area.

5.1 U.S. DOT Connected Vehicle Research Program

The auto manufacturer interviews addressed their perspectives on the connected vehicle research program. Overall, the auto manufacturers interviewed voiced that Crash Avoidance Metrics Partnership (CAMP) was effectively leading technical development and that Vehicle Infrastructure Integration Consortium (VII-C) was effectively leading policy development. Those interviewed felt that the Safety Pilot Model Deployment is an important step and that the 2013 target for NHTSA decision has helped create and sustain momentum for connected vehicle technology (CVT).

Figure 5-1: Crash Avoidance Metrics Partnership Vehicle Safety Communications Consortium (VSCC) members *(Source: National Highway Traffic Safety Administration 2013)*

Connected vehicle technology development requires Original Equipment Manufacturers (OEMs) to cooperate to an unprecedented extent; however there are several challenges regarding standards, security, privacy, liability and governance that need to be resolved.

The participants voiced perspectives on standards and security, governance, privacy, liability, harmonization, implementation path and competitive technologies, wireless spectrum and GPS, and U.S. DOT role in Vehicle-to-Infrastructure (V2I).

Standards and Security
Regarding standards, technical development is further along than policy development, and standards policy development still needs to be addressed. In particular, the questions of how many OEMs adopt the standards and whether they adopt them in total or only partially need to be resolved. Security remains a difficult issue and is yet to be resolved. A less sophisticated solution for managing certificates is acceptable initially, but as the number of vehicles grows, the solution will need to be more robust.

Figure 5-2: Connected vehicle environment
(Source: U.S. DOT 2013)

Governance
OEMs are concerned about the cost of establishing and operating a governance structure since there seems to be no sound business case for a private sector entity to assume governance responsibility. If the U.S. DOT or NHTSA were a candidate to manage governance, concerns include whether existing statutory authority is adequate, funding of the governance entity, and whether a government agency could assume a long term operational commitment. Although NHTSA leadership is very much aware of the governance issue and is actively trying to develop a solution, it may be difficult because NHTSA has no authority regarding infrastructure.

Privacy and Liability
OEMs expressed that there will likely be public concern about the use of connected vehicle technology data regarding location and speed for law enforcement purposes or for advertising. However, if the public is well educated in the benefits of CVT, they may be willing to make trade-offs with regards to privacy.

Regarding liability, OEMs acknowledge that litigation may be inevitable and is a cost of doing business. However, there is a greater concern when a vehicle uses V2V data that comes from another manufacture's vehicle. The receiving vehicle is relying on the quality of the sending vehicle's CVT, thus exposing their vehicles to additional liabilities.

Harmonization
Regarding harmonization, all noted that the U.S. is on a path closer to Europe using 5.9GHz while Japan is using 700MHz spectrum. U.S. harmonization with Canada is very important and desired long term with Mexico. It was noted that the focus on crash imminent safety applications is unique to the U.S., while Europe and Japan are focused on mobility applications.

Implementation Path and Competitive Technologies
For OEMs, a market-driven implementation is preferable to a mandated regulation. The absence of a benefit or value proposition for the initial connected vehicle owners is a major issue; therefore, companies may defer implementation, leaving only the regulatory approach. In addition, competitive technologies using radar, cameras and on-board sensors have emerged offering immediate benefits that do not require other vehicles to be similarly equipped. The OEMs expressed that if a regulatory approach is used, the regulations need to be flexible enough to avoid stifling further technological development.

5 Automotive Manufacturer Perspective

Figure 5-3: Examples of connected vehicle competitive technologies *(Source from left to right: Think Stock 2013; courtesy of Volvo 2013; Steve Jurvetson/Wikimedia Commons/Public Domain; derivative work: Mariordo. 2013)*

Wireless Spectrum and GPS

When asked about a current requirement for the National Telecommunications and Information Administration (NTIA) to examine the possibility for spectrum sharing of the relevant 5.9GHz spectrum, the OEMs considered spectrum sharing an important risk, owing to the possibility of harmful interference. The threat of such a situation would result in the implementation of CVT being endangered.

Regarding the evolution of GPS, backward equipment compatibility was considered critical. It would be economically impractical to upgrade CVT systems already on the road for changes in GPS.

U.S. DOT Role in V2I

The U.S. DOT may be able to help accelerate CVT by supporting mobility/V2I applications and model deployments. Residual infrastructure from model deployments could help create value for early CVT vehicle owners. The challenges include investment by local and state governments, multiple competing wireless paths into the vehicle (e.g. cellular, Bluetooth, WiFi, and satellite radio), and determining who would process the data in V2I/mobility applications.

5.2 Considerations for Moving Forward to a Connected Vehicle Environment

Specific recommendations from OEMs resulting from the interviews are summarized below.

1. *Support resolution of governance issues.* Governance issues need to be addressed to support the goals of the OEM community. OEMs have practical implementation concerns regarding resolution of security issues and privacy policy as well as how adherence to standards will be enforced and how evolution of standards will be managed in the future. Governance is the mechanism that both allows and ensures competitors are able to successfully collaborate - not only among themselves but also with others to bring to fruition the promise of connected vehicle technology.

2. *Help prepare the public.* The OEMs are conscious that vehicle purchasers early in the rollout of CVT will realize few benefits. The public needs to be educated prior to implementation to better understand the benefits and privacy implications of the technology. In addition, law enforcement and legislators need to fully understand CVT in order to regulate data use that supports the full benefits of the technology.

3. ***Secure the future with government support for foundational elements.*** The GPS system and 5.9 GHz spectrum allocation are foundational elements for connected vehicle technology. Automotive hardware has an unusually long life with limited or no ability to upgrade it. Therefore, the GPS system must be committed to supporting, without degradation, legacy automotive hardware. Similarly, it is imperative that adequate spectrum be allocated for CVT. The OEMs are concerned that the failure of either of these would compromise the success rollout of the technology.

4. ***Recognize that competitive technologies may cause consumer and OEM resistance to a CVT rollout.*** New camera and sensor based technologies that offer some of the benefits of CVT are increasingly available as options or even standard equipment. It is important to establish a compelling vision for the public of the unique and longer term benefits of building a connected vehicle environment. The availability of these new technologies, in conjunction with the incremental price paid for connected vehicle technology, may discourage investment in connected vehicle equipment.

6 Post-Hoc Analyses

This section focuses on the post-hoc analyses, conducted to assess the longer-term impacts and benefits of investment in ITS. The goal of this effort was to examine how the performance of various systems have changed over time either due to expansion/enhancement of the systems, or changing traffic patterns or traveler behavior. Evaluations include assessment of:

- a transit traveler information system in Portland, Oregon
- a ramp metering deployment in Kansas City,
- high occupancy toll (HOT) lanes in Minneapolis/St. Paul, and
- an arterial management system in Phoenix, Arizona.

Overall, the post-hoc evaluations demonstrated:

- Archived performance data from areas with sufficient data quality procedures can be used to support a data analysis to review trends over time, without the need to collect more field data.
- Ongoing and expanded ITS implementations continue to produce measurable effects. In some cases, initial benefits can be sustained over time. In other cases, the benefits may change or decrease as the system infrastructure and operational strategies, evolve but the implementation can still produce a positive benefit-cost ratio, especially if incremental operational costs decline.
- Advances in technology, such as Portland's transit traveler information system which provided 511 transit arrival times over mobile devices, can have a disruptive effect on benefits.

Three of these analyses were carried out in time horizons with fairly stable, built-out network environments, and for these analyses, results show meaningful trends or implications. The single evaluation in Phoenix, Arizona with significant variations in population and demand during the analysis period proved more complex to evaluate and interpret.

Each of the four case studies provided meaningful insights on the longer-term impacts from the implementation of ITS technologies and systems. The key outcomes from each study along with the questions answered by each evaluation are summarized below.

The Portland Transit Tracker – Traveler information dynamic message boards along the Barbur Boulevard corridor bus stops coincided with a 6 percent increase in transit ridership for this corridor while regional ridership was nearly unchanged from 2001 to 2005. In 2005, mobile 511 traveler information made the dynamic messages ubiquitously available. Ridership along the Barbur corridor from 2005 through 2012 declined to parallel regional ridership trends.

> The initial ridership increase from bus stop transit traveler information was mitigated by more global provision of this data through telephone, internet and mobile devices

Figure 6-1: Traveler Information via mobile device
(Source: Think Stock 2013)

- Did message boards in Barbur corridor increased ridership? **Yes, until ubiquitous 511**
- Was Barbur ridership significantly different from regional trends? **Yes, until ubiquitous 511**

The Kansas City I-435 Ramp Metering – The implementation of ramp metering in 2010, and evaluation from 2010-2011 suggested that overall volume increased by as much as 20 percent. The travel speed and volumes through this corridor during the period from 2010 through 2012 during the peak period varied slightly, but this variation was not statistically significant. Further, incident clearance time savings were also maintained from the initial implementation with slight variations that were not statistically significant. Consequently, the benefits from ramp metering were maintained over the two years, post implementation.

Figure 6-2: Ramp metering *(Source: Wikimedia Commons/Patriarca12 2013)*

To answer the key questions posed by this post-hoc analysis:

- Has the increased throughput in the corridor remained stable? **Yes**
- Have the speed impacts in the corridor remained stable? **Yes**
- Have the reduced incident clearance times remained stable? **Yes**

The MnPASS I-394 HOT lanes and I-35W Expansion – The 6 percent increase in travel speed and 5 percent increase in volume reported during peak periods in the six-month period after I-394 HOT lane implementation in November 2006 have been maintained through year 2012. Expansion of the system to I-35W in late 2009 required a capital investment that was 37 percent that of the initial I-394 investment with an operating cost increase of only 17 percent above I-394 operating costs. Benefits from the I-35W expansion were observed in a 3 percent increase in corridor volume while maintaining consistent speed on the mainline and HOT lane. Overall cost per transponder transaction has declined by 32 percent from year 2006 to year 2011. So while benefits are slightly lower, the lower operating costs may result in a higher overall benefit-cost ratio than for the initial implementation.

Figure 6-3: MnPass HOT Lanes Pass *(Courtesy Minnesota DOT 2013)*

- *Have the initial impacts of the system on I-394 changed over time?* **No, benefits maintained**
- *Did the expansion to I-35W result in similar benefits?* **No, slightly lower benefits**
- *Have economies of scale been obtained by I-35W expansion?* **Yes, 32% lower cost/transaction**

The Scottsdale Road, Phoenix Arterial Traffic Signal Coordination – The 2001 implementation resulted in an average speed increase of 6.2 percent among other benefits; however, the result should not be compared against the 2001 findings given differences in data types. This region exhibited huge fluctuations in travel demand and multiple changes to the system. Comparison of performance of the Scottsdale Road facility with a non-coordinated arterial suggests the Scottsdale Road facility exhibited similar volume variations with far lower fluctuations in speed. Scottsdale Road corridor was successfully adapted to widely fluctuating demand levels. The traffic signal coordination improvements may have provided more stability in travel speeds and times relative to the other corridor that was not similarly equipped.

Figure 6-4: Traffic Signal Control *(Source: Think Stock/David De Lossy 2013)*

6 Post-Hoc Analysis

- *Did volume increase along the corridor remain steady over time* — No, volume fluctuated
- *Did speed increase along the corridor remain steady over time* — No, speed fluctuated
- *Are there continued benefits from the traffic signal coordination* — Yes, lower fluctuations in speed

This work provides confidence that continued operations and maintenance on ITS deployments is a prudent use of public funds, generating sustained benefits over time. The work also demonstrates the need to factor in the cost side when comparing potential investments as incremental costs are often reduced as the system expands. The effort offers a glimpse into performance-based management, showing that the performance of the system can be monitored over time using archived data and the findings used to make adjustments in order to optimize the return on transportation investments. Finally, this work points to the need for public sector agencies to continually monitor technology trends for their potential impact on transportation systems, as technology advances can have a disruptive effect on the benefits of an older technology.

A separate report on the post-hoc data analyses is available for more information on this aspect of the Longitudinal Study of Implementation: *ITS JPO Longitudinal Study: Post-Hoc Data Analyses of ITS Deployments in Portland, Kansas City, Minneapolis-St. Paul, and Phoenix* (US DOT, March 2013).

7 Public Workshop Summary

The public workshop occurred as a special session of the Chicago ITS Industry Forum on *Connected Vehicles: Moving from Research Towards Implementation* on September 27, 2012. This session brought together 75 participants representing state and local transportation agencies, automotive and device manufacturers, and consultants, along with representatives from the U.S. DOT, Noblis, and professional organizations such as ITS America, ATRI, and AASHTO. Among this group were eleven public sector staff who participated in the Longitudinal Study of ITS Implementation and were invited to attend by the ITS Joint Program Office. Input from this session was intended to be considered as the U.S. DOT shapes its approach to supporting investment in the connected vehicle environment of the future.

7.1 Workshop Objectives

The objectives of the workshop were to:

- Discuss experiences, challenges, and factors influencing the decision to adopt ITS technologies in the past and implications for connected vehicle technologies.
- Interpret the value proposition for connected vehicle from the stakeholder perspective.
- Identify ways to overcome barriers for investing in connected vehicle technologies.

The desired outcomes of the workshop were:

- A better understanding of how to convey benefits and costs information and measures of effectiveness for decision-makers.
- Suggested additional actions the USDOT should consider to assist public agencies in creating the connected vehicle environment.

7.2 Word Cloud

During a morning exercise, the participants were asked to write down "three words I'd like to share about connected vehicles." The word cloud shown in Figure 7-1 was created from the audience's responses.

Figure 7-1: Connected Vehicle Word Cloud *(Source: Noblis 2013)*

The overwhelming choice of the participants was the word "Safety," followed by "Mobility," which indicates that the messages regarding the use of connected vehicle technology for safety and mobility are being heard. Additional word choices such as "Future," "Opportunity," "Transformative," and "Exciting," reflect the energy and enthusiasm of the industry for the connected vehicle environment. However, the words, "Challenging," "Complex" and "Complicated" indicate that enthusiasm is tempered by the sense that significant barriers must be overcome so that the program can move forward. The words, "Liability," "Integration," "Interoperability," and "Security" reflect some of the issues that must be resolved.

7.3 Key Themes

The participants were asked to describe barriers to connected vehicle implementation and discuss ways their organizations could overcome these barriers. The following themes emerged from the discussions throughout the day:

1. **Accelerate the timeline.** There is a strong sense that connected vehicle environment is going to happen, as shown by the responses to the first of the poll questions. (Over 80 percent of respondents rated the probability of a connected vehicle environment being established in the next 15 years as greater than 50 percent.) However, commenters noted that the U.S. DOT's timeline may not adequately account for the rapid pace of technological development of device manufacturers and automobile manufacturers. There is concern that these entities may move ahead to implement connected vehicle applications in a different manner than what is envisioned by the U.S. DOT research program. A number of comments on the word wall echoed the sentiment, "Just do it."

2. **Focus on upcoming policy decisions.** The U.S. DOT decision regarding the 5.9 GHz spectrum allocation and NHTSA decisions stemming from the safety pilot will have a substantial impact on the growth of connected vehicle environment. From the discussion, it

appears that these decisions will shape the direction of this emerging market for the next few years. There are a number of stakeholders who are waiting on these decisions before they make investments in connected vehicle technology. Some commenters stated that there are potential connected vehicle players, such as insurance companies, who might not be aware of the program but will be impacted by these policy decisions.

3. **Recognize the need for "just right" amount of governance.** There was general consensus that achieving the connected vehicle vision will require more cooperation among state and local transportation agencies, automobile manufacturers, communications providers, device providers, and insurers, among others. As the presenters noted, these organizations have cooperated well in the past on initiatives such as the 511 Coalition, but there may need to be new structures to support the connected vehicle research effort. The issue of governance was a recurring theme throughout the discussion, with commenters recommending some mechanism to set and assure the use of standards and interoperability, but many expressing wariness about government mandates. The challenge of balancing these goals will be ongoing for the connected vehicle research program.

4. **Demonstrate the value proposition with measurable benefits**. Multiple presenters and commenters emphasized that an effective description of the value provided by new technology is a major factor in past ITS deployment decisions and is needed to move forward with connected vehicle technology. Many of the speakers and participants highlighted the need to define the value to the end user of participating in a connected vehicle environment. This can be accomplished though economic measures but also through compelling narratives about the safety, mobility and sustainability improvements made possible by connected vehicle technology. In describing the value to automobile manufacturers, the commercial vehicle industry, and state and local agencies charged with infrastructure investment, commenters were nearly unanimous about the need to define economic impact of connected vehicle technology using quantitative measures such as Benefit-Cost, ROI, and payback periods. The NCHRP 101-03 study currently being conducted will be useful for Vehicle-to-Infrastructure (V2I) investment decisions, but commenters agreed that more information will be needed.

5. **Leverage funding models.** There was discussion about the funding models available to the state and local agencies that will be charged with implementing the connected vehicle infrastructure, with commenters noting the overall decline in federal highway spending. A few commenters suggested that connected vehicle technology could be used to generate revenue so that the infrastructure projects are self-sustaining, while others remarked that there is little public or industry support for that. A representative from the tolling industry remarked that his organization has had successful public-private partnerships in infrastructure deployment projects and that over time public support grows as they realize the benefits of a shorter commute.

6. **Increase awareness of state and local agency staff, decision makers, and the public**. Multiple commenters expressed the view that there is lack of awareness of the connected vehicle program among state and local agencies that will deploy it, decision makers who will make the investments, and the general public. A commenter pointed to the safety pilot segment on the *Today Show* as a good first start for increasing awareness. Another commenter noted that marketing efforts should have a clear, focused message that is repeated through multiple media channels.

7. **Start building the workforce.** A smaller number of commenters expressed concern about the lack of expertise needed to implement the connected vehicle infrastructure. There is a need for a mix of computer science, communications, transportation engineering, and business aspects to successfully implement connected vehicle projects. Others expressed the view that these capabilities are available in the current workforce, but that these people may be employed outside the public sector.

7.4 Participant Response

The Longitudinal Study of ITS Implementation participants expressed their appreciation for the opportunity to learn more about the connected vehicle research program, to interact with their peers, and to share their perspectives and concerns on ITS deployment and the connected vehicle research program. The following is typical of the messages received:

"Thank you for a good meeting with a lot of interesting discussions and presentations. The meeting has prompted some additional reviews of my systems to better determine how we can be ready for and actively participate to move the connected vehicle Initiative forward."

Figure 7-2: Interactive Exercise - Connected Vehicle Word Wall *(Source: Noblis 2013)*

8 Considerations for Next Generation ITS

The goal of this study was to understand the factors influencing public agencies, automobile manufacturers, and the commercial vehicle industry decisions related to the adoption, growth, replacement, or cancellation of ITS technologies and systems to inform the implementation of the next generation ITS and the connected vehicle environment. Our interviews and workshops with the public sector transportation agencies, the trucking industry, and automobile manufacturers found that these stakeholders have both overlapping and differing decision factors and concerns regarding the connected vehicle program and other ITS investments as shown in Table 8-1.

Table 8-1: Stakeholder Top Decision Factors and Key Questions regarding Connected Vehicle Technology *(Source: Noblis 2013)*

Stakeholder	Top Decision Factors	Key Questions
State and Local Transportation Agencies	1. Budget/Funding Sources 2. Quality and Reliability 3. Stakeholders/Partners on Board 4. Agency priorities 5. Demonstrable Benefits	• What is our role in the connected vehicle research program? How can I get involved? • What government support will be provided to implementing agencies in terms of funding and technical assistance? • What benefits should we expect from V2I connected vehicle implementation? Does this present a valid business case to support investment? • How strong is federal government commitment to implementing this technology? • How quickly will connected vehicle technology be adopted in vehicles?
Trucking Industry	1. Return on Investment 2. Compatibility with existing systems	• How much will this technology cost vs. its expected benefit? • Is the technology mature enough for implementation? • Will the technology be compatible with my existing systems? • Will privacy of company and driver data be protected?
Automobile Manufacturers	Consumer/Market demand	• Will consumers accept this technology? • How much are they willing to pay for it? • Will government mandate OBEs? • What government support will be provided to ensure spectrum and develop communication standards? • What is OEM liability for driver or government use of connected vehicle data?

In order for the connected vehicle environment to flourish, the value proposition needs to be made clear to each party and the key questions answered. Once stakeholders are convinced of the value proposition from their perspectives, it will be possible to discuss how to coordinate decisions among the key parties. Such decisions would be in the area of standardization, timing, funding, and evolutionary deployment plans.

8.1 Applying Past Experience to Achieve Future Success

One of the premises of this effort was that learning about the past decision processes associated with ITS implementation could yield insights that would encourage positive outcomes from decision-making regarding ITS deployment and next generation ITS. While the type of technology being evaluated may change, it is reasonable to assume that key decision factors will remain fairly stable for various stakeholder groups.

Comparing the key decision factors identified through the interview task with the qualitative information needs cited for pursuing connected vehicle applications validates this premise. The top five decision factors for the public sector presented in Table 8-1 were all mentioned or emphasized in one way or another during the discussion of connected vehicle technologies in the interview process. For example, the concern raised about an unfunded mandate correlates with the importance of identifying budget/funding sources in making ITS implementation decisions. Since connected vehicle applications are closest to the initiation phase in the implementation lifecycle, the most important factors stakeholders will consider at this time are those that ranked higher in the initiation phase: budget/funding sources, demonstration of benefits, involvement of stakeholders and partners, and agency priorities (see Figure 3-5). As testing and early deployment of these technologies and applications proceeds, key stakeholder concerns will shift more to the factors rated higher in the development and deployment phases such as quality and reliability of the products, end-user awareness and acceptance, and knowledge and expertise available to be able to operate and maintain the system.

The themes listed below contain recommendations for addressing these key decision factors.

8.2 Cross-cutting Themes

Overall, cross-cutting analysis of the stakeholder interviewers, post-hoc data analysis, and workshop reveals several major themes for the federal government to consider regarding next generation ITS and the connected vehicle environment:

1. *Clearly define and publicize benefits for connected vehicle to engage stakeholder interest.* Many public organizations are unclear of what the connected vehicle program is and what the benefit of implementation would be. In particular, several state and local stakeholders in Georgia and Arizona are questioning the business model of implementing these new systems.

2. *Recognize that private sector prefers a market driven approach on the vehicle side while public sector seeks stronger federal guidance on infrastructure deployment.* The private sector feels that a market based approach should drive connected vehicle implementation, however public agencies are asking for guidance from federal government

with respect to applications that they would implement, their role in the connected vehicle program, standards to ensure interoperability and federal funding. There is a concern that connected vehicle applications will be mandated without federal funding to back them up.

3. **Ensure demonstrations include diverse constituents.** Demonstrations of connected vehicle technologies should include diverse constituents in terms of modality, levels of congestion, and size of deployment to establish a robust peer group for market share growth. Some public stakeholders do not know how to get involved and feel like they have missed the boat. For example, a western state transportation staff member expressed that they far less likely to receive grants for connected vehicle deployments because they are a rural area. In addition, decision makers in New Jersey feel like they have missed their opportunity to participate in connected vehicle deployments since they were not among the first test sites and now need guidance for how to get involved.

4. **Focus on education and information dissemination.** Across the board, this study showed the need for education to inform the public sector, trucking industry, and end users about connected vehicle technologies and the benefits they can achieve. The ITS Professional Capacity Building (PCB) programs, AASHTO, APTA, and other media were suggested as means to spread the word about the future connected vehicle environment.

5. **Support resolution of governance issues including security, privacy, and adherence to standards** Governance issues need to be addressed to support the OEM community in moving forward. OEMs have practical implementation concerns regarding resolution of security issues and privacy policy as well as how adherence to standards will be enforced and how evolution of standards will be managed in the future. Governance is the mechanism that both allows and insures competitors are able to successfully collaborate - not only among themselves but also with others to bring to fruition the promise of connected vehicle technology.

6. **Secure the future with supporting commitments from other U.S. Government agencies.** The GPS system and 5.9 GHz spectrum allocation are foundational elements for connected vehicle technology. Automotive hardware has an unusually long life with limited or no ability to upgrade it. Therefore, the GPS system must be committed to supporting, without degradation, legacy automotive hardware. Similarly, it is imperative that adequate spectrum be allocated for CVT. The OEMs are concerned that the failure of either of these would compromise the success rollout of the technology.

7. **Recognize that competing technologies will temper consumer, trucking industry and OEM enthusiasm for a connected vehicle technology rollout.** New camera and sensor based technologies that offer some of the benefits of CVT are increasingly available as options or even standard equipment. It is important to establish a compelling vision for the public of the unique and longer term benefits of building a connected vehicle environment. The availability of these new technologies in conjunction with the incremental price paid for connected vehicle technology, may discourage investment in connected vehicle equipment

8. **Reduce the likelihood for long-term risk aversion by establishing incremental successes with connected vehicle pilots and demonstrations.** Risk for connected vehicle implementation is a concern for all stakeholders, but especially state and local agencies. Shortcomings of connected vehicle implementation could result in pushing innovators and early adopters toward the late majority for technology acquisition.

9. ***Consider ways to provide federal support for continued operations and maintenance of existing and future ITS infrastructure and systems.*** The post hoc analysis showed that ITS benefits from deployed systems are maintained pretty consistently over time. Expansions do not always offer the same benefits as initial deployment, but economies of scale usually drive down the costs of these expansions. In any case, providing funding to support continued operations and maintenance is generally a wise investment. As the installed base of ITS has increased, agencies are increasingly concerned with the quality, reliability, maintainability of the systems. Barriers to funding ongoing operations and maintenance activities are not helpful as agencies consider how best to manage their systems.

10. ***Define national guidelines for connected vehicle implementation*** to minimize incompatible or duplicate systems from being developed and to ensure a consistent deployment approach. The public sector is looking for national leadership in the connected vehicle environment, with deployment guidance as well as in setting standards. The decision making process is more complex, with many more actors in this environment, and states are concerned about unfunded mandates. The public sector is also looking for the business case and is concerned about who would pay for road side infrastructure in a connected vehicle environment. For companies such as long-haul trucking companies that operate across the U.S., the need to purchase several redundant technologies would be a significant burden.

8.3 Areas for Future Research

Using both quantitative and qualitative analysis, this study identified key factors and their relative importance in the decision to adopt ITS and in subsequent decisions to grow, replace, contract, or cancel technologies or systems. The study also surfaced important areas for future research:

- **The effect of disruptive technology.** The post-hoc analysis of the Barbur Boulevard project in Portland, OR indicated that advances in technology, such as the availability of traveler information over mobile devices, can have a disruptive effect on benefits from an ITS deployment. How public sector agencies can anticipate and plan for these technology advances is an important area of inquiry.
- **End user views of connected vehicle technology.** There is a fourth stakeholder whose views have not been included in this study due its research scope--the end-user of connected vehicle technology. The Connected Vehicle Safety Pilot Program currently underway in Ann Arbor, MI will test performance, evaluate human factors and usability, observe policies and processes, and collect empirical data to present a more accurate, detailed understanding of the potential safety benefits of connected vehicle technologies in a real-world implementation. The data from this pilot will be critical to supporting the 2013 NHTSA agency decision on vehicle communications for safety and future decisions on connected vehicle technologies.

References

1. Amodei, Richard, Erin Bard, Bruce Brong, Frank Cahoon, Keith Jasper, Keith. "Atlanta NAVIGATOR Case Study." Atlanta, Georgia. November 1, 1998a.

2. U.S. Department of Transportation. "Review of Existing Literature and Deployment Tracking Surveys: Decision Factors Influencing ITS Adoption" 2012
   http://ntl.bts.gov/lib/45000/45600/45616/FHWA-JPO-12-043_v2_Final_508.pdf

3. U.S. Department of Transportation, ITS Joint Program Office. "ITS Strategic Research Plan, 2010 – 2014 Progress Update 2012"
   http://www.its.dot.gov/strategicplan/pdf/ITS%20Strategic%20Plan%20Update%202012.pdf

APPENDIX A. List of Project Reports

Task 2
- U.S. Department of Transportation. (2012) *Review of Existing Literature and Deployment Tracking Surveys: Decision Factors Influencing ITS Adoption*.
  http://ntl.bts.gov/lib/45000/45600/45616/FHWA-JPO-12-043_v2_Final_508.pdf

Task 3
- Webinar: *Why Do Transportation Agencies Adopt ITS? Join the Conversation and Share your experiences with the ITS Evaluation Program*, June 7, 2012.
  http://www.pcb.its.dot.gov/t3/s120607_its_eval.asp
- Site Interview Summaries:
  - Idaho
  - Georgia
  - Phoenix, Arizona
  - Tucson, Arizona
  - Washington
  - Maryland/New Jersey
  - Trucking Industry
  - Automobile Manufacturers
- Task 3 Summary Report: *Longitudinal Study of ITS Implementation: Survey Findings*

Task 4
- Individual Post-hoc Analyses:
  - Transit Traveler Information – Barbur Boulevard, Portland
  - Ramp Metering – I-435 Corridor, Kansas City
  - HOT Lanes —I-394 and I-35W, Minneapolis/St. Paul
  - Arterial Traffic Signal Coordination– Scottsdale Road, Phoenix
- Task 4 Summary Report: *Longitudinal Study of ITS Implementation: Post-hoc Analyses*

Task 5
- Workshop Summary Notes

Task 6
- Final Summary publication: *Longitudinal Study of Implementation: Decision Factors and Effects*

APPENDIX B. List of Acronyms

AASHTO	American Association of State Highway and Transportation Officials
ACHD	Ada County Highway District
ADOT	Arizona Department of Transportation
AMBER	America's Missing: Broadcasting Emergency Response
APTA	American Public Transportation Association
APTS	Advanced Public Transportation System
ARC	Atlanta Regional Council
ATIS	Advanced Traveler Information Systems
ATA	American Trucking Association
ATM	Advanced Traffic Management
ATMS	Advanced Traffic Management System
ATRI	American Transportation Research Institute
AVL	Automatic Vehicle Location
BCDLL	Benefits, Costs, Deployment, Lessons Learned
BMC	Baltimore Metropolitan Council
CAD	Computer Aided Dispatch
CAMP	Crash Avoidance Metrics Partnership
CCTV	Circuit Television
CEO	Chief Executive Officer
CHART	Coordinated Highway Action Response Team
COG	Council of Governments
COMPASS	Community Planning Association of Southwest Idaho
CVISN	Commercial Vehicle Information System and Network
CVO	Commercial Vehicle Operations
CVT	Connected Vehicle Technology
DMS	Dynamic Message Signs
DOI	Diffusion of Innovations
DOT	Department of Transportation
DSRC	Dedicated Short Range Communications
EOBR	Electronic Onboard Recorder
ETC	Electronic Toll Collection
EMS	Emergency Medical Services
FCC	Federal Communications Commission
FHWA	Federal Highway Administration
FMCSA	Federal Motor Carrier Safety Administration
FTA	Federal Transit Administration
GDOT	Georgia Department of Transportation

U.S. Department of Transportation, Research and Innovative Technology Administration
Intelligent Transportation System Joint Program Office

APPENDIX B. List of Acronyms

GIS	Geographical Information System
GPS	Global Positioning System
GRTA	Georgia Regional Transportation Authority
HOT	High Occupancy Toll
ICM	Integrated Corridor Management
IHW	Idaho Health and Welfare
ISP	Idaho State Police
IT	Information Technology
ITS	Intelligent Transportation Systems
JPO	Joint Program Office
LSI	Longitudinal Study of Implementation
MAG	Maricopa Association of Governments
MAP	Metropolitan Atlanta Performance
MAP-21	Moving Ahead for Progress in the 21st Century
MARTA	Metropolitan Atlanta Rapid Transit Authority
MCDOT	Maricopa County Department of Transportation
MDOT	Maryland Department of Transportation
MDSP	Maryland State Police
MDTA	Maryland Transportation Authority
MPO	Metropolitan Planning Organization
MTA	Maryland Transit Administration
NCHRP	National Cooperative Highway Research Program
NHTSA	National Highway Traffic Safety Administration
NJDOT	New Jersey Department of Transportation
OEM	Original Equipment Manufacturer
PAG	Pima Association of Governments
PCB	Professional Capacity Building
PDA	Personal Digital Assistant
R&D	Research and Development
RITA	Research and Innovative Technology Administration
RFP	Request for Proposal
RM	Ramp Meter
ROI	Return on Investment
SCC	State Communication Center
SHA	State Highway Administration
SO&M	System Operations and Maintenance
SRTA	State Road and Tollway Authority
TMC	Transportation Management Center
TMS	Traffic Management System

APPENDIX B. List of Acronyms

USDOT	U.S. Department of Transportation
VMS	Variable Message Sign
VII-C	Vehicle Infrastructure Integration Consortium
VSCC	Vehicle Safety Communications Consortium
WSDOT	Washington State Department of Transportation

APPENDIX C. Site Visit Summaries

Georgia

Figure C-1: Transportation Management Centers in Georgia *(Source: Noblis 2012)*

Noblis conducted ten in-person interviews, a panel session, and three facilities tours during the Georgia site visit. Interviews ranged from 1.5 to 2.0 hours and included ITS decision-makers from key state, regional, and local agencies. This site visit was unique in that the team was able to capture the perspectives of a more isolated agency in a smaller, suburban area. Specifically, Johns Creek is a more insular organization given it geographic distance from high density, large metropolitan areas. Their perspectives did yield a different perspective compared to other site visits.

Participating agencies included Georgia Department of Transportation (GDOT), Gwinnett County Department of Transportation, Georgia State Road and Tollway Authority (SRTA), City of Atlanta, City of Johns Creek, and the Georgia Regional Transportation Authority (GRTA). Among these stakeholders, Noblis gathered freeway, arterial, planning, transit and tolling perspectives related to the adoption, growth, and replacement of ITS systems. Facility tours highlighted the operations of three traffic management centers (TMCs) operated by Gwinnett County, SRTA, and GDOT.

Notable themes expressed among the panel session and individual interviews are listed below:

- *Funding challenges now and into the foreseeable future are a reality for transportation agencies within Georgia.* A 1% sales tax for transportation funding in the state of Georgia was put on the ballot for taxpayer voting in a July 31, 2012 referendum. Economists projected that if approved statewide, this one percent sales tax would generate $18.67 Billion over the 10-year life of the tax to fund regional transportation projects across Georgia. The increase in sales tax was rejected by voters in 9 of 12 regions (including Atlanta) and did not pass. Consequently, transportation improvement funding will be very limited throughout most of the state, including the Atlanta region.

- *The Atlanta region had varying opinions on the impact of funding challenges.* On the one hand, less money in the transportation program implies the competition will be fierce for the project dollars, and ITS projects must compete in this environment. On the other hand, since there will be less funding for new highway and transit system capacity expansions, the highest priority becomes properly maintaining the system you've got and then operating it as efficiently and safely as you can. In this context, ITS projects get more priority since they can be a cheaper solution and help to maximize the efficiency of the system.

- *Funding limitations may lead to agencies deploying systems that are not interoperable.* An implication of the funding shortfall is variation in funding across jurisdictions. This variation can lead to silo implementations that may not work with neighboring systems. These will likely generate greater challenges in the future when attempting to establish interjurisdictional coordination.

- *GDOT is clearly a leader in the adoption and implementation of ITS technologies within the region.* GDOT began implementing ITS technologies before the 1996 Olympics and many transportation agencies in the region look to GDOT for help with specifications on projects. A major consideration for agencies when implementing an ITS project is maintaining interoperability with GDOT operations. For example all of the surrounding areas are adopting the NaviGAtor 2.0 ATMS to enable interoperability with GDOT and other local agencies.

- *A desire for interoperable systems has created new cross jurisdictional and institutional issues.* Several agencies mentioned their on-going activities to create interoperable systems and specifically the ability to share video. However, IT departments have been reluctant to open up internal networks to other agencies. Each agency and it's respective IT department has been protective of their networks and very concerned with network security. These institutional issues need to be resolved in order to have fully interoperable systems.

- *The Atlanta region is well positioned for performance monitoring and already moving in the direction of performance-based management, but agency stakeholders expressed several questions regarding what will be required by the federal government.* Staff from the Georgia Regional Transportation Authority (GRTA) conduct performance monitoring on an ongoing basis and prepares an annual performance monitoring report for the Atlanta region, the latest being the 2010 Transportation Metropolitan Atlanta Performance (MAP) Report[1]. The 2010 Transportation MAP Report updates the annual Transportation MAP Report, which sets performance measures for tracking the performance of the transportation system in Metropolitan Atlanta. Measures are organized in six general categories—Mobility, Transit Accessibility, Air Quality, Safety, Customer Satisfaction, and Transportation System Performance. In the future, the Atlanta Regional Council (ARC) will be responsible for implementation of the new performance-based management legislative provisions associated with MAP-21 (in the Atlanta region). ARC has administered the congestion management system within the region, so has some experience in this regard, which should be useful for implementing the performance-based management requirements. In an effort to gear up for it's new role, over the past year, ARC has been collecting and analyzing field data that hasn't been historically collected by planning agencies in the Atlanta region. GRTA and ARC frequently share information and results relating to transportation performance evaluation methods, and a collaborative relationship has formed between the staff of these organizations.

---

[1] http://www.grta.org/tran_map/2010_Transportation_MAP_Report.pdf?bcsi_scan_cd8b447107cc943b=1

Appendix C. Site Visit Summaries

- *Decision makers were unsure what the public sector's role will be in the connected vehicle research program as well as the eventual implementation period.* Agencies can see the value and are very interested in connected vehicle applications but are more aware of the V2V applications that can be largely handled by the auto manufacturers. Several of the decision makers were not very aware of the V2I applications. They would l ke more demos and information on what is planned for V2I and business cases to support justification as well as describing what their role would be.

- *The structure of the ITS decision-making process within the Atlanta region has evolved and been mainstreamed and streamlined into the transportation planning process, but the overall complexity has grown in several ways.* There are many more agencies and stakeholders making ITS decisions in the region now, compared to when the region first implemented ITS in preparation for the 1996 Olympics. As a result, more deployment decisions need to be coordinated and there are sometimes potential overlaps in roles and responsibilities that must be clarified. Also, the benefits of systems being considered are evaluated more closely now prior to making deployment decisions. Regional initiatives have been started to encourage the sharing of resources and avoiding needless duplication, and technical users groups are formed as needed. For example, a NaviGAtors users group meets periodically to discuss related issues and best practices. The City of Atlanta is leveraging public-private partnership opportunities and partnerships with other organizations to accomplish goals at lower cost.

- *It is very difficult to estimate the benefits of enabling and foundational technology deployments, such as traffic monitoring devices, cameras, and DMS signs.* Some stakeholders suggested that minimum standards may need to be established for items such as the density of cameras per mile. The details of what would be required of them from the Section 1201 rule or the specific language contained in MAP-21 were not well known by the regional stakeholders. They expressed a concern that a change the infrastructure may be required to accommodate gathering more data for performance measures. In any event, Atlanta transportation stakeholders realize that some amount of money will be required to collect the necessary data. In tight economic times, transportation agencies simply do not have the money to conduct a post-hoc analysis on the performance of the project or system, which could then be compared to the original estimates for benefits

Idaho

Figure C-2: Transportation Agencies in Idaho *(Source: Noblis 2012)*

Appendix C. Site Visit Summaries

Noblis conducted six in-person interviews, a panel session, and two facilities tours during the Idaho site visit. Interviews and the panel session ranged from 1.5 to 2.0 hours and included ITS decision-makers from key regional and state stakeholders including the Idaho Transportation Department (ITD), the Ada County Highway District (ACHD), the Idaho State Police (ISP), the Idaho Department of Health and Welfare (IHW), and Community Planning Association of Southwest Idaho (COMPASS). Among these stakeholders, Noblis achieved arterial, highway, transit, EMS, and police perspectives related to the adoption, growth, and replacement of ITS systems. Facilities tours highlighted the operations of the Idaho State Communication Center (SCC) and the ACHD Traffic Management Center.

Notable themes expressed among the panel session and individual interviews are listed below:

- *Integration and interoperability of different systems is very important within each agency, between agencies, and with neighboring states.* For example, ACHD TMC routinely shares their cameras with the SCC. Also, ITD is ensuring consistency of 511 message sets with states along the Northwest Passage. Further, ITD owns and operates DMS within Washington State where it borders Idaho. ITD, ACHD, ISP, and IHW emphasized that integration of technologies and requirements for new systems to work with existing systems are key factors in the decisions to adopt new systems or upgrade existing systems.
- *Idaho agencies' designation of peers clearly differentiates themselves from larger metropolitan regions such as Seattle, Minneapolis, Washington, and such.* The ISP noted that their peers would be state such as Wyoming or Montana. They turn to these peers when determining the 'right fit' for deployments of technologies. The ISP, ITD, and ACHD have at times looked to larger parallel organizations in other regions when considering replacement of systems or adoption of new systems, particularly for cases where peers do not have experience to share.
- *Idaho agencies within the Boise region view themselves ranging from innovators to early majority compared to their peers, while other organizations away from this area are viewed as late adopters.* The SCC is the first of its kind to house a state police dispatch as well as provide emergency dispatch and communications for EMS, Idaho Transportation Department, hazardous material incidents, public health emergencies, AMBER Alerts and many other situations. The ISP noted that their organization was an innovator in 2000 and served as a beta-tester for a CAD system; however, the CAD system did not meet expectations. Consequently they are now an organization that is an "early majority" compared to peer state police organizations. Surrounding regional entities look to ACHD as the innovator, and these entities were categorized as "late majority adopter" of ITS technologies.
- *Agencies both at the state, regional, and local levels need better, clearer information on what performance metrics can most effectively be measured to manage by performance.* The measures that are appropriate for highly congested, high demand regions will be different than for rural environments. Often for small implementations measures may not prove meaningful because there is little to no congestion in general. For example, a safety system implemented at a rural intersection will have 'noise' that is potentially greater than the effect of the system.
- *Noted was the lack of documented benefits for some ITS technologies specific to their regional characteristics.* The transit perspective as well as the highway perspective noted the absence of quantified benefits data for rural environments. Some noted the need for

Appendix C. Site Visit Summaries

expansion of the ITS Benefits, Cost, and Evaluation database to include more rural and transit data points.

- *In implementing ITS systems, the lack of expertise in-house and through outsourcing as well as rapid changes in technologies often delays or defers adoption or expansion of technologies.* An ITD led project to deploy APTS across a number of rural transit agencies was delayed multiple years given the changing state of certain vehicle-tracking technologies. Additionally ITD, ACHD, and ISP noted the need for development of expertise in-house for deployment and operations of ITS systems. For example, ISP trained in-house staff on the use of GIS with their CAD system.
- *Many organizations have undergone restructuring that has introduced new upper management that for some have supported ITS and for others have made ITS deployment funding and justification more challenging.* The key theme among restructuring is performance-based management and demonstration of benefits. The ACHD's new management required significant education on the operational benefits from ITS before supporting ITS investments, while restructuring in ITD's highway operations group has brought management that are challenging ITS investment.

Maryland

Figure C-3: Maryland Coordinated Highway Action Response Team (CHART) Facilities *(Source: Noblis 2012)*

The metropolitan Washington region is unique in that two states (Virginia and Maryland) as well as the District of Columbia come together for coordinated action. Within this region exists a number of densely populated cities and counties along with suburban and rural constituencies. Noblis conducted a panel session along with eight in-person or on-the-phone interviews and a facility tour during the Maryland Department of Transportation (MDOT), State Highway Administration (SHA) site visit. Interviews and the panel session ranged from 1.5 to 2.0 hours and included ITS decision-makers from key regional and state stakeholders:

- MDOT SHA Coordinated Highways Action Response Team (SHA CHART),
- MDOT Maryland Transportation Authority (MDTA)
- MDOT Maryland Transit Administration (MTA),
- Maryland State Police (MDSP),
- Baltimore Metropolitan Council (BMC),
- Metropolitan Washington Council of Governments (COG),
- Montgomery County Transportation Department.

Appendix C. Site Visit Summaries

Among these stakeholders, Noblis gathered arterial, highway, transit, EMS, and police perspectives related to the adoption, growth, and replacement of ITS systems. The facility tours highlighted the operations of the MD SHA CHART Traffic Management Center.

Notable themes expressed among the panel session and individual interviews are listed below:

- *Although there is value in quantifying accomplishments, in some instances ITS may not lend well to performance metrics.* There needs to be a balance between performance based metrics and other activities that cannot be measured but are equally effective. Many expressed concern over how to evaluate ITS technologies implemented at a smaller scale. Montgomery County DOT is unclear as to what metrics should be collected for their transit signal priority and is looking for guidance. The Maryland State Police liaison to CHART also expresses interest in establishing the right metrics for their support to CHART and would benefit from guidance.
- *There has been a shift from implementation to M&O and outcomes.* With a focus on performance measurement and a constrained funding environment, decision makers have shifted their focus toward the maintenance and operation of their ITS systems. The SHA explained that they had significant systems roll out in the 1990s and again recently with their 511 system. The focus for them is small incremental operations improvements with their existing systems over the adoption of larger new ITS systems.
- *Regional interaction is an important consideration in ITS deployment decisions.* In the 90s, ITS technology was deployed to "try things out" with no consideration for regional interaction. However, in the last 5 to 7 years, the whole concept has changed; new implementations follow the regional architecture and ensure that all stakeholders are on board.
- *Federal requirements are making it difficult for local jurisdictions to apply for federal funding.* Local jurisdictions have a difficult time applying for federal funding because of the requirements associated with the funding.
- *Although there has been a decrease in overall transportation funding, the transit industry may experience growth.* Because of the current economic constrains, transit may see an increase in ridership. With the increases in ridership are increased expectation of services and implementation of technologies that can meet this higher expectation.
- *There is value in connected vehicle technology, but state and local agencies are waiting for national direction and clarity, particularly in the role of the infrastructure.* There needs to be more direction and information on the role and expectations of the state and local agencies. In addition, the menu of applications is too large and would be better to see more focused implementations nationwide.

Phoenix, Arizona

Figure C-4: Transportation Agencies in Phoenix, Arizona *(Source: Noblis 2012)*

Noblis conducted six in-person interviews, a panel session, and three facilities tours during the Phoenix-Scottsdale site visit. Interviews and the panel session ranged from 1.5 to 2.0 hours and included ITS decision-makers from key regional, state and local agencies. The participating agencies included Arizona Department of Transportation (ADOT), City of Scottsdale Department of Transportation, City of Phoenix Department of Transportation, Arizona Department of Public Safety, Maricopa Association of Governments (MAG), and Maricopa County DOT. Among these stakeholders, Noblis gathered freeway, arterial, planning and police perspectives related to the adoption, growth, and replacement of ITS systems. Facilities tours highlighted the operations of the Maricopa County Traffic Management Center, Scottsdale Traffic Management Center and ADOT Traffic Management Center.

Notable themes expressed among the panel session and individual interviews are listed below:

- *Importance of a regional organization that brings together all of the ITS agencies.* The AZTech regional operations group brings together 25 agencies in the Phoenix Metro Area. This group has been in place since 1996 and has made the Phoenix Metro Area take a regional approach to transportation operations. AZTech also has defined regional performance measures published on their website.
- *Importance of a good relationship with the local FHWA representative.* ADOTs and other local agencies relationship with FHWA has allowed them develop and expand their ITS systems well beyond what they would be able today without support from FHWA.
- *Shift in focus to operation and maintenance after a lot of deployments in the past years.* Maricopa County expressed that for years the region has been focused on implementing and deploying new ITS technologies. Now the region is at a point where they have to begin operating and maintaining their systems with maximum efficiency. This has led agencies to become more involved in creating regional and cross jurisdictional operational plans early on in the project lifecycle.
- *Concern on what is required from state and local entities for collection of data and analysis for performance management.* All participating agencies expressed a concern of what data will be required to fulfill any federal performance measurement requirements. An even bigger concern is that the agencies do not have the staff or expertise to do detailed analysis. Simple performance measures have already been important in the

region for showing benefits of ITS technologies and securing funding for new projects. But most agencies expressed that they have more data available but cannot justify additional data analysis in tight operating budgets.

- *There is a need for federal requirements to determine what data needs to be collected from different ITS projects.* Requirements for data collection and analysis need to be determined by the federal government. This should include who is responsible for the performance management analysis, for example the MPO. MAG already has one full time staff member dedicated to performance monitoring and they believe a third party handling performance analysis may be ideal.
- *ITS technologies have the potential as a "force multiplier" for benefits.* Benefits of ITS projects may go well beyond the typical benefits of congestion mitigation or other traffic measures. For example, installing CCTV cameras for traffic monitoring helps the police be more efficient as well. TMC operators can locate incidents before police can and provide location information. TMC operators can also implement operational strategies (DMS messages or others) to warn approaching motorists of the incident. These actions help prevent secondary incidents and allow the police to spend less time responding to incidents and more time doing other work. Performance measures should reflect this added benefit.
- *Encouraging in-house innovation is cost effective and keeps innovation as a priority.* With a good in-house technical staff, ADOT has encouraged in-house development of technologies such as their wireless communications and copper theft warning system. The in-house development of projects has saved ADOT money, and allowed them to be innovators in the ITS industry.
- *Maricopa County is a connected vehicle test bed location.* MCDOT and its partners are testing multiple priority/preemption applications for transit and fire vehicles. MCDOT is working on policy and institutional issues so they can begin installing permanent in-vehicle devices in fire and transit vehicles to begin collecting data. One of the major overall concerns is security, including who will maintain the security system. Fleet vehicles (especially police, fire, and EMS) are a great way to begin testing, promote awareness, and gain public acceptance of connected vehicle technologies.

Tucson, Arizona

Figure C-5: Traffic Management Centers in Tucson, AZ *(Source: Noblis 2012)*

Appendix C. Site Visit Summaries

Noblis conducted four in-person interviews with five people, a panel session, and two facilities tours during the Tucson site visit. Interviews and the panel session ranged from 1.5 to 2.0 hours and included ITS decision-makers from key regional, state and local agencies. The participating agencies included Arizona Department of Transportation (ADOT), City of Tucson Department of Transportation, Pima Association of Governments (PAG), and Sun Tran. Among these stakeholders, Noblis gathered freeway, arterial, planning and transit perspectives related to the adoption, growth, and replacement of ITS systems. Facilities tours highlighted the operations of the Tucson Traffic Management Center and the local ADOT Traffic Management Center.

Notable themes expressed among the panel session and individual interviews are listed below:

- *The rapid changing of technology can be a barrier to adoption.* Technology is changing very rapidly and there is a fear that the technology will become obsolete in five years. The future maintenance of technology being deployed today is a concern for all agencies. Michael Hicks, ITS Manager for the City of Tucson, stated that one of the most important factors that he considers when making ITS Technology decisions is if the technology will be viable and maintainable in 5 to 10 years.
- *For ITS decision makers, it is critical to have an understanding of more than just ITS.* Decision makers must know IT, communications technology, and other technology trends. ITS technologies today are driven by network and communications technologies. It is also critical to work closely with IT departments to facilitate the connection of ITS and communications and network technology.
- *Concern on what is required from state and local entities for collection of data and analysis for performance management.* All participating agencies expressed a concern of what data will be required to fulfill any federal performance measurement requirements. An even bigger concern is that the agencies do not have the staff or expertise to do detailed analysis. Simple performance measures have already been important in the region for showing benefits of ITS technologies and securing funding for new projects. But most agencies expressed that they have more data available but cannot justify additional data analysis in tight operating budgets.
- *In a constrained economy, sharing resources has been important to continuing ITS growth.* For example in the Tucson region, the transportation department, fire department, and police department all share access to the Regional Transportation Data Network (RTDN). This allows all of the agencies to share maintenance and upgrade costs while receiving the same benefits. Inter-agency cooperation has saved money and allowed agencies to continue to build ITS systems.
- *Performance measures have the potential to be the bridge between technical and non-technical decision makers.* It is difficult to quantify the difference in benefits between an ITS strategy such as a signal timing plan or a traditional engineering solution such as adding a lane. Performance measures can fill this gap in knowledge. For example, if an agency invests in ITS strategies they would see 'X' benefits but if they do another project they see 'Y'. The manner in which these benefits can be explained is critical to getting non-technical decision makers to understand the benefits between projects. Performance measures will hopefully become easier allowing more agencies to analyze performance and have a solid basis for justifying ITS projects.

Appendix C. Site Visit Summaries

Washington State and New Jersey

Noblis conducted four additional interviews in person and over the phone. The additional interviews consisted of transportation stakeholders from New Jersey and Washington State. Interviews ranged from 1.5 to 2.0 hours and included ITS decision-makers from state and transit agencies. The participating agencies included New Jersey Department of Transportation (NJDOT), Washington State Department of Transportation (WSDOT), and King County Metro. Among these stakeholders, Noblis gathered freeway, arterial, and transit perspectives related to the adoption, growth, and replacement of ITS systems.

Notable themes expressed among the individual interviews are listed below:

- *Performance reporting can be very effective but some agencies lack staff and expertise.* WSDOT has seen clear advantages with their performance reporting but reports that it is costly to conduct performance monitoring even with a good system already in place. WSDOT reports that performance monitoring has been effective but it is difficult to measure the performance of some ITS technologies such as dynamic message signs. NJDOT expressed institutional and cultural barriers to performance reporting that they are actively trying to overcome. Some agencies lack the skills and staff, while others lack the willingness to change their ways of operating to accommodate performance data collection and analysis.

- *ITS has proven to be a low cost-high benefit option but faces tough competition.* ITS technologies have shown to be a low cost and high benefit solution. However, state agencies expressed the challenge that politicians want to deliver high profile construction projects and ribbon cuttings. The investment in ITS also becomes a lifelong investment in operations and maintenance which is often at the mercy of the software developers.

- *State agencies expressed issues involving using federal funding because of the 'strings' that come attached.* NJDOT returned a large sum of money after they were unable to use the funding for their project. They had $13 million for the construction of a new adaptive signal control system but were not permitted to use any of the funds to do the design of the system. NJDOT could not obtain design funds and therefore was not able to use the construction funds.

- *Multimodal integration is of high priority to state and local agencies.* King County lost a federal grant to upgrade their signal priority system and integrate into the City of Seattle's ITS network but the decision was made to continue the project with only state funding because of the high benefits. It was expressed that multimodal integration could be easier if FTA and FHWA worked closer together. Transit agencies would like to see transit more integrated with FHWA, especially in regards to funding. One interviewee noted that if a transit agency has FTA money and wants to work with an agency that has FHWA money, they cannot work together.

- *A small transit industry reduces competition and willingness to produce well developed standards.* Limited competition in the transit industry leads to fewer options for creation of transit agencies systems. Each private company has its own proprietary system that makes interoperability difficult. It was expressed that when the transit industry does get involved in a standard it is usually only to ensure that the standard will fit the products they already have.

- *Education and training opportunities offered by local division federal representatives' focus on infrastructure and would prove more valuable had they greater expertise on ITS and operations.* Many state agencies have expressed that they have shifted focus to operations and need help to operate their ITS systems with greater efficiency. While some interviewed agencies expressed significant value from their federal counterpart's expertise and assistance, NJDOT interviewees suggest federal representatives would prove more helpful had they a stronger ITS and operations knowledge and experience.

- *Several concerns were expressed with the connected vehicle program.* Risk and liability with connected vehicle technology has been an area of concern. Specifically, King County Metro expressed concern about having public and private vehicles sharing a connected vehicle network with two very different needs and security levels. At the State and local levels, there seems to be a general understanding of the V2V applications, however a clear need was stated for information and understanding of V2I applications. In addition, New Jersey DOT also expressed interest in joining connected vehicle research programs but feel like they missed their opportunity. It would be beneficial for state and local parties to be aware of opportunities to become involved in the connected vehicle program.

APPENDIX D. Interview Instruments

Public Sector Pre/Post Webinar Survey

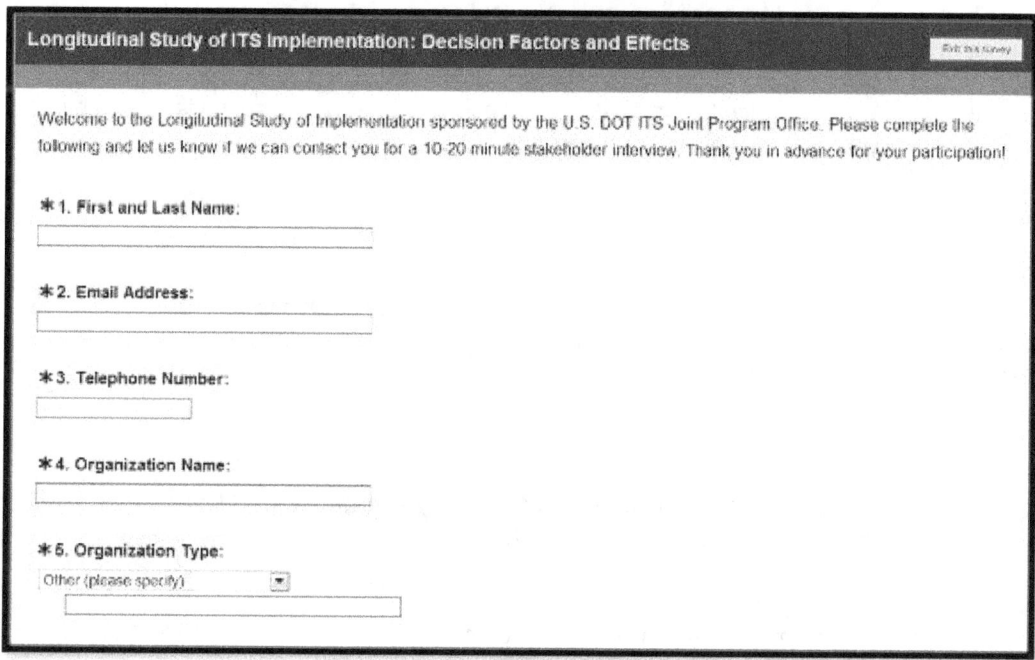

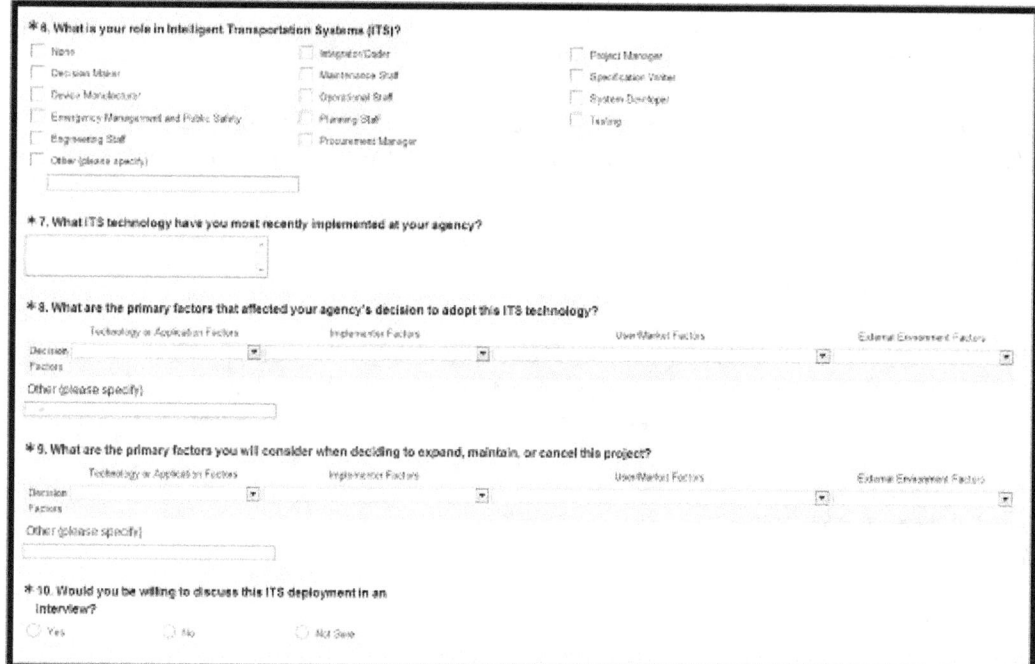

Appendix D. Interview Instruments

Public Sector Screening Interview form

May I please speak to _____?

My name is _____ and I am calling from Noblis. We are a non-profit science, technology, and strategy organization that advises the government on transportation issues. We are conducting research on behalf of the U.S. DOT on ITS adoption and implementation. At an earlier time you indicated that you would be willing to participate in a 20 minute survey on your experiences with ITS decisions made by your organization. Is this a convenient time to continue?
1 – Yes   go to Background
2 – No   go to Better time

Better Time
The interview would last about 20 minutes, and can be arranged for a time convenient to your schedule. Is there another time we could contact you?
1 – Yes   schedule appointment
2- No   Thank you for your time

Reluctant
I understand, I'll let you go. But if you would, are there others you might suggest that are able to share information about your organizations ITS decisions? Thanks for your help.

Background
As you know, involvement in this interview is entirely voluntary. You may decline to answer any of the interview questions you do not wish to answer and may terminate the interview at any time. All information you provide will be considered confidential. We estimate this survey will take no more than 20 minutes. The purpose of the study is to investigate factors that influence an organization's decision to adopt ITS technologies, and their decision to grow, replace, or cancel ITS technologies. The data collected will be used by the U.S. DOT to evaluate options for promoting ITS adoption. Individual respondents will be not identified by name unless you give express permission. Are you ready to continue?
1 – Yes   go to begin survey
2 - No   go to better time
3 – wants more info   go to details

Details
 If you have any questions regarding this study, or would like additional information to assist you in reaching a decision about participation, please feel free to contact the James Pol with the U.S. DOT ITS Joint Program Office (JPO) at *James.Pol@dot.gov* or call 202-366-4374.
Are you ready to continue?
1- Yes   go to begin survey
2 – No   go to better time

Appendix D. Interview Instruments

Begin survey
I will begin the survey now.

1.0 Demographic Information
Confirm the following:

1.1 Name will be confirmed in introduction.

1.2 Organization:

1.3 Type of Organization:

1.4 What position do you have (at above)? Position in Organization:

2.0 Technology Adoption

2.1 In managing a transportation system, organizations make decision to adopt, grow, replace, or cancel specific ITS technologies. Thinking about this, what was your organization's most recent **major** ITS decision?

2.2 What technology or application was being considered? What was the scale of this deployment/implementation?

2.3 When was the funding allocated for this implementation?

2.4 What was your role in the decision making process? (Is there a formal decision making process your organization follows?)

2.5 In your opinion, was this a good decision?

2.6 Can you describe the problem you were trying to solve?

3.0 Decision Factors

3.1 What were the key factors that influenced this decision? (*Provide if hesitant*, price, funding, met regulatory requirements, solved congestion or safety problem)

3.2 Were there external factors that influenced this decision, for example, neighboring or peer agency experience, economic environment?

3.3 What type of analysis, if any, was used to select this technology? (*Provide if hesitant* e.g. benefit cost study, case study, site visit to neighboring agency, bid process, other)

3.4 What additional resources would be helpful during your decision making process? (BCDLL Knowledge Resource Database, Training on Tech, other training, standards information, contact info on prior deployers, guidance documents/guides)

Thank you, you've been very helpful thus far, we have just 4-5 more questions to cover.

Appendix D. Interview Instruments

4.0 Organization Information

4.1 Is there an individual or group with specific ITS planning/funding responsibility or expertise within your organization? If yes can you elaborate? If no, where do you get ITS expertise?

4.2 How would you describe your organization's attitude toward new technology? (1=Innovator and 10=Last Adopter)

4.3 Are there other ITS decisions that your organization is considering or has recently made about specific technologies? And if yes, can you describe the top 3 decisions (adopt, grow, maintain, replace, cancel) and the technologies being considered?

4.4 Are there other organizations in your region (for example, an EMS department or a Transit agency) that have recently made an ITS decision (adopt, grow, maintain, replace, cancel)?

Interviewer Instructions

If you feel this person was 'interesting' and is a good candidate for further discussion, then read below. Ask yourself, would we benefit from having another hour of this person's time? If no, go to closing.

The next step for our study is to conduct in depth site visits and conduct a stakeholder workshop with organizations like yours that have unique insight gained through multiple ITS decisions. Can we contact you in 4-6 weeks about possible participation in these events?

4.5 *Optional (see rating below)*: Would you be interested in participating in a stakeholder interview at your location?

If YES, Thank you very much. We will be contacting you within a week to provide information about the process for setting up a site visit. Do you have any questions/comments?

Closing
Thank you for your participation. We appreciate your help in informing the ITS deployment program. If you have additional thoughts, please feel free to share them with us through my email address (*provide analyst's email address*).

Interviewer Rating of interviewee when considering asking for follow up interview:					
Size of deployment	1	2	3	4	5 (small – large)
Recentness of deployment	1	2	3	4	5 (old – recent)
Depth of information provided during interview	1	2	3	4	5 (sparse – detailed)
Good information about peers/intrajurisdiction	1	2	3	4	5 (limited – detailed)

Appendix C. Interview Instruments

Public Sector In-Depth Interview Form

1.0 Interviewee/Organization Information

Name:
Title:
Organization:
Years in current position:
Years at current organization:
Years working in this region:
Years in transportation industry:

Primary responsibilities (please describe):

Technology Adoption Continuum – According to Geoffrey Moore in *Crossing the Chasm* a gap exists from one to the next adopter group. This gap must be crossed if a technology or innovation is to continue to be adopted. "Innovators" are the first to adopt a technology while "Early Adopters" are looking for radical "change agents" to give them an advantage and become leaders. The Early Majority on the other hand are focused on productivity improvements to improve existing operations and do not want to disrupt their organization. The "Late Majority" adopts technology once it's well established and benefits are clear. In contrast, laggards are the last to adopt and are averse to change. While a gap exists from one to the next group of adopters, there exists a chasm between "Early Adopters" and the "Early Majority" and is the result of how each views the purpose behind innovation and considers their peers. For successful technology adoption, this chasm must be crossed.

Where along the technology adoption continuum would you place?
 a. Yourself with regard to personal technology acquisitions?
 b. Yourself with regard to your work environment?
 c. Your organization compared to regional peer organizations?
 d. Your region compared to national peer organizations?

Has your organization recently undergone a reorganization? If yes, what was the impetus for this reorg? Also, how does this reorg affect your responsibilities?

Appendix C. Interview Instruments

2.0 Mapping the Current ITS Decision Process

Please describe a recent completed ITS technology decision. Was this a decision to adopt, grow, maintain, cancel, or replace a technology or system? What was the time frame for this decision? What was the 'problem' being solved?

2.1 Please describe/sketch out the decision making process that your organization followed to reach this decision. (Use as many papers as needed.)
    Sample phases of decision process:
        Initiation – Development – Implementation
        System justification – Implementation - Use
    problem definition - needs assessment – system evaluation/cost justification – submission to decision makers (e.g. board of directors, finance division, etc), bid – purchase – replacement/ customization – integration, education on use and benefits, training, ongoing support and maintenance
    Decision making process sketch, for example, to involve MPO, long range plans, ops divisions, studies, etc.

2.2 Who were the key personnel (titles) involved in this decision and what role(s) did they play?

2.3 Is this the typical decision-making process for investments, and if not how did this differ from the typical?

2.4 What was the economic environment at the time this decision was made?

2.5 What federal or state resources supported this decision, or were absent but would have supported this specific decision?

2.6 Rate the importance of factors during the initiation, development, and implementation phases of the technology adoption process. Focus on individual and relative ratings.
Rating scale: 1= unimportant, 10=critical

Initiation Period - *Areas of attention in the initiation period include: the initial consideration and meetings and generation of ideas, shocks and triggers that initiated the development of an innovation, and the development of plans and budgets*

Development Period - *Areas of attention in the development period include shifting of performance criteria, top management involvement and roles, altering relationships and cooperation among stakeholders, confronting setbacks, and fluid participation of personnel. Activities include evaluation of ideas, design of innovation, development/evaluation/modifications to prototype.*

Implementation Period - *Areas of attention in the implementation process are the deployment of implementation of the technology or practice, linking the new with the old (systems, processes), developing user expertise, or even early termination of the innovation.*

Appendix C. Interview Instruments

Factors	Sub-category	Importance of Factors/Subcategories (1-10)		
		Initiation	Development	Implementation
Technology/ Application	price			
	readiness and maturity			
	quality and reliability			
	demonstration of benefits			
	sufficiently robust standards			
	open source/non-proprietary software			
	interoperability mapped			
Implementing Organization	perception of risk			
	knowledge and expertise available			
	clarity in division of responsibilities			
	partners on board			
	organizational priorities			
	adoption by peers			
User/ Market	the role of interest groups			
	end-user acceptance			
	end-user awareness/understanding of benefits			
External	local, regional, state, federal political support			
	budget/funding sources			
	overarching agency priorities			
	presence of a regional architecture			
	involvement in the project by stakeholders			
	external champions			
	links with universities and research centers			

Appendix C. Interview Instruments

2.7 Rate the importance of specific barriers during the initiation, development, and implementation phases of the technology adoption process. Focus on individual and relative ratings.
Rating scale: 1= unimportant, 10=critical

Barriers	Sub-category	Importance of Factors/Subcategories (1-10)		
		Initiation	Development and Spread	Implementation
Available information (knowledge)	Lack of information on knowledge resources			
	Lack of information on markets			
	Lack of qualified personnel			
Technical	Lack of interoperability			
	Lack of standardization and certification			
	Difficult adaptation to a new technology			
Legal and regulatory	Legislation, regulations, taxation			
	Administrative barriers			
	Legal risk			
Financial and Economic	High costs (financial and other)			
	Lack of funds within the enterprise and subsidies from outside			
	Lack of peers in the market			
Cultural and Societal	Low acceptability			
	Poor attitude of personnel towards change			
	Inability to devote staff to innovation activity			
Decision-making	Lack of cooperation among partners (public, private,...)			
	Fragmentation of decision levels			
	Lack of Vision and Policy Growth			

2.8 Were there systems that were considered for adoption but not selected, or systems that were implemented but subsequently mitigated or cancelled? Why was it not selected, mitigated or cancelled?

Considered but not selected:

Implemented then mitigated/cancelled:

Appendix C. Interview Instruments

3.0 Evolution of the ITS Decision Making Process - 1990 - 2000 – 2010 and beyond

How has your ITS decision making process changed from the 1990's to 2000 to 2010 and beyond? Has the complexity of the decision environment increased? Are you interacting with new players or facing new contracting paradigms?

4.0 Role of Performance Management and ITS

What do you think is the impact of the new performance-based management legislation on ITS decision-making in the next 5 years? Does your organization currently operate a performance management system? Does the presence of a performance monitoring system *currently* enable you to better justify funding for projects? Can you provide any examples?

5.0 Factors Influencing Prioritization in a Constrained Economic Environment/Benefits Profile

Has there been a recent period, and do you foresee in the future, an environment with greater economic constraints for transportation funding? What have you observed, or what do you perceive will be likely trade-off in this environment competing with ITS projects? What factors will support or detract from ITS prioritization? Has your organization tracked whether/how ITS benefits change over time? What benefits data points are available for your ITS systems from a historical, repetitive, or current perspective?

6.0 Focus on Interoperability and Integration

What are the challenges your organization faces with regard to systems interoperability, proprietary equipment, legacy systems, and such. Are these challenges faced with expertise and in-house skills development or through contracting?

7.0 Your Thoughts and Awareness of the Connected Vehicle

Can you tell us what you know about the US DOT connected vehicle initiative? Are you familiar with any specific envisioned applications? How is this program relevant to your region? What are the top 3 unknowns about this initiative? What are some strategies that would prove most effective in promoting awareness of and interest in the connected vehicle initiative?

Appendix C. Interview Instruments

Trucking Industry Interview Form

1.0 Demographic Information (to be completed before the interview)

1.1 Company description (include if applicable: number of power units, sector operating in, and commodities hauled)

1.2 Title/Position of Interviewee:

2.0 Technology Adoption

2.1 What was the last major technology decision that your company made? (Remind the interviewee that this could include choosing to install, increase, upgrade or remove a technology.)

2.2 What problem or issue was being solved?

2.3 How many units were installed, upgraded or removed?

2.4 What was the per-unit cost of this technology?

2.5 Please describe the decision making process that your company followed to reach the decision to install, upgrade or remove this specific technology.

    2.5.1. Who were the key personnel (titles) involved in this decision and what role(s) did they play? Is this a typical arrangement for similar technology decisions?

    2.5.2. What were the key factors that influenced this decision? (*Prompt interviewee if necessary:* price, funding, to meet regulatory requirements, to solve safety or productivity problem.)

    2.5.3. Based on the answers provided, ask interviewee to rank order the decision factors, with 1 being the most important.

2.6 What was your role in the decision making process?

2.7 What were the positive outcomes of the decision?

2.8 What were the negative outcomes of the decision?

3.0 Decision Factors

3.1 Please describe how the following factors affected the decision to select this technology or application:

    3.1.1. The price of the technology.

    3.1.2. The readiness and maturity (proven quality and reliability; market penetration) of the technology.

    3.1.3. That the technology has been shown to provide clear benefits.

    3.1.4. That the technology is compatible with other previously installed systems.

3.2 What type of analysis, if any, was used to select this technology or application? (*Prompt interviewee if necessary: benefit cost study, case study, bid process, ROI, payback period.*)

Appendix C. Interview Instruments

3.3 What additional resources would have been helpful during your decision making process? *(Prompt interviewee if necessary: training on technology or other training, info on standards, contact info for prior users, guidance docs, user guides.)*

4.0 Organization/Technology Planning Information

4.1 Is there an individual or group with specific technology planning/funding responsibility or expertise within your organization? If yes, can you elaborate? If no, where do you get technology expertise?

4.2 How would you describe your organization's attitude toward new technology in general? (1=Innovator and 10=Last Adopter)

4.3 For the following organizational objectives that may impact technology adoption, please rate the importance of each on a scale of 1 to 5 (with 1 being extremely important and 5 not being important at all).

	1	2	3	4	5
LOGISTICS					
Manage driver efficiency	❏	❏	❏	❏	❏
Improve dispatching	❏	❏	❏	❏	❏
Optimize fleet utilization	❏	❏	❏	❏	❏
Improve on time performance	❏	❏	❏	❏	❏
Automate vehicle location tracking	❏	❏	❏	❏	❏
Identify stolen vehicles quickly	❏	❏	❏	❏	❏
MAINTENANCE					
Automate and manage service/repair orders	❏	❏	❏	❏	❏
Accelerate vehicle repair & maintenance	❏	❏	❏	❏	❏
Reduce emissions	❏	❏	❏	❏	❏
COST					
Reduce fuel consumption	❏	❏	❏	❏	❏
Reduce vehicle maintenance costs	❏	❏	❏	❏	❏
Reduce insurance premiums	❏	❏	❏	❏	❏
Automate road/toll collections	❏	❏	❏	❏	❏
SECURITY & SAFETY					
Improve vehicle security/safety	❏	❏	❏	❏	❏
Improve driver security/safety	❏	❏	❏	❏	❏
Improve cargo security/safety	❏	❏	❏	❏	❏
Reduce traffic violations	❏	❏	❏	❏	❏
CUSTOMER RELATIONS					
Improve customer relationship management	❏	❏	❏	❏	❏
Assist or automate warranty management claims	❏	❏	❏	❏	❏
Automate payments or transactions for delivered goods	❏	❏	❏	❏	❏

Appendix C. Interview Instruments

4.4 For each of the following technologies, please indicate whether your company already has it, is likely to add it in the future or does not plan to add it:
(*Note any technologies previously discussed by interviewee before asking)

	Already have it	Will add within 3 years	Considering adding it in the future	Considered but decided not to add	No plans to consider
LOGISTICS/MANAGEMENT TECHNOLOGIES					
Vehicle tracking	❏	❏	❏	❏	❏
Electronic access to client / cargo / order information from vehicle	❏	❏	❏	❏	❏
Satellite / cellular-based communication between terminal and vehicle	❏	❏	❏	❏	❏
SAFETY TECHNOLOGIES					
Load stability sensors	❏	❏	❏	❏	❏
Lane departure warning systems	❏	❏	❏	❏	❏
Forward collision warning systems	❏	❏	❏	❏	❏
Roll stability systems	❏	❏	❏	❏	❏
Remote diagnostic system that senses malfunctions in vehicle and notifies driver or carrier	❏	❏			❏
Ability to link employee's cell phone or PDA to the vehicle	❏	❏	❏	❏	❏
Adaptive cruise control	❏	❏	❏	❏	❏
Infrared cameras that can project the image of objects in the road onto a display	❏	❏	❏	❏	❏
Acoustic/visual parking aid	❏	❏	❏	❏	❏
Back-up cameras/audible warnings	❏	❏	❏	❏	❏
NAVIGATIONAL TECHNOLOGIES					
Built-in satellite navigation system (GPS)	❏	❏	❏	❏	❏
Hand-held navigation systems	❏	❏	❏	❏	❏
Real-time, on-demand traffic information	❏	❏	❏	❏	❏

4.5 Are there any other technologies that your company has added or plans to add in the future? If yes, please describe.

4.6 Are there any other technologies that your company has considered but ultimately decided not to add? If yes, please describe.

5.0 Public Sector Involvement

5.1 What do you think are the top things that the public sector currently does well to support ITS deployment in the trucking industry?

5.2 What do you think are the top things that the public sector could do better to support ITS deployment trucking industry?

6.0 The Connected Vehicle program

Appendix C. Interview Instruments

6.1 Are you aware of the U.S. DOT's Connected Vehicle program?

    6.1.1. If yes, please describe the technologies and/or services that you are familiar with, as well how you learned about the program.

        6.1.1.1. Do you think your organization would be likely to purchase these services in the future?

        6.1.1.2. What kind of information do you believe your organization would require to encourage investment in these services?

        6.1.1.3. What strategies do you think would be effective in promoting awareness of and interest in the connected vehicle initiative?

If no, which communications or media channels would you most prefer to learn about the program?

Appendix E. Workshop Summary

APPENDIX E. Webinar Summary

As part of the preparation for the detailed public sector site visits and interviews, Noblis developed and conducted a webinar that provided an introduction and overview of this study. The webinar was also designed to gain interactive feedback on specific factors influencing ITS decision-making, and to elicit additional interest in participating in the interview process. The development of this webinar included close participation and input from USDOT representatives.

Webinar Objectives and Audience

The webinar titled, "Why do transportation agencies adopt ITS? Join the conversation and share your experiences with the ITS Evaluation Program" was sponsored by the ITS Professional Capacity Building Program's Talking Transportation Technology (T3) webinar series and held on June 7, 2012. The webinar presentation slides are provided in several formats through the ITS PCB Program's T3 Archives at http://www.pcb.its.dot.gov/t3_archives.aspx.

The webinar was hosted with three objectives relevant to Task 3:
- Share results of Study of Implementation literature review surfacing key factors influencing the decision to adopt ITS.
- Confirm key factors influencing an agency to adopt ITS and identify the key factors leading an agency to expand, maintain, contract or cancel ITS project.
- Invite participants to deepen involvement through a screening interview.

The target audience for the webinar was anyone who has deployed or is considering adoption of ITS technologies and those interested in discussing and learning about the factors that affect ITS deployment. Participant counts varied during the webinar with an average of 80 participants throughout. Webinar attendees included consultant (21%), state and local DOT employees (20%), federal employees (9%), transit agency employees (8%), academic (5%), planning agency employees (4%), public safety employee (1%), and other (4%). These results are presented in figure to the right. Those that chose other and provided written input to the webinar Q&A box included individuals from turnpke authorities and local environmental regulatory agencies.

Figure E-6: Webinar participant affiliation
(Source: Noblis 2013)

During the webinar registration process, participants were asked to provide examples of their ITS implementations to share during the discussion. Several participants provided examples from their experience in the field. While some asked to keep their names and agencies confidential, others were willing to share publicly their successes and challenges. These real world examples were introduced in the webinar to share the results of the literature review and explain the factors that go into ITS decision making. The webinar was interactive in nature, including asking eight survey questions:

- One question about participants role in transportation,
- Four questions, each one requesting participants to select the most important decision factor among the four sets when considering adoption of an ITS technology.
- Three questions, each one requesting the most important decisions factor when making an ITS decisions to grow, maintain, and contract/cancel a system or technology.

Overall, participants confirmed that the factors proposed in the literature review are the right set capturing the array of factors influence the decision to adopt, as well as expand, maintain, contract or cancel ITS.

Factors Influencing ITS Adoption

The first set of polls administered live during the webinar asked participants to recollect a recent *ITS adoption decision* and then select the most important factors influencing this decision within each of the four factor categories –technology or application, implementer, external environment, and user/market. Given the survey mechanism and limits on the number of factors visually communicated, some of the factors are combined or placed in an "other" category. A total of up to six factors were presented, with the capacity to select a single factor. Given the diverse population of webinar participants, not all had recently participated in an ITS decision; consequently sample size ranged between 46 and 58 responses among the four polls on selecting the most important factor for their ITS adoption decision. The frequency with which specific factors were selected by participants as most important is presented in the charts below for the four factor categories.

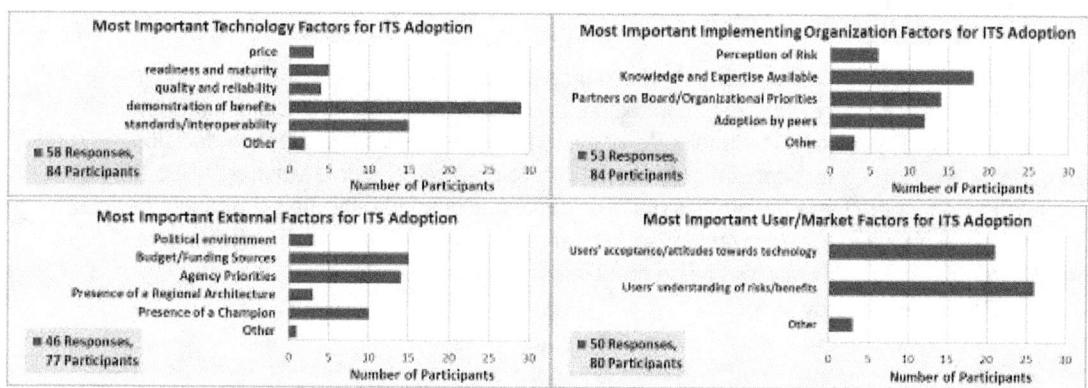

Figure E-7: Factors Cited as Most Important When Deciding to Adopt an ITS Technology or System *(Source: Noblis 2013)*

An overwhelming number of webinar participants (50%) cited demonstration of benefits as most important among technology, followed by standards/interoperability (25%). For far fewer participants, price, readiness and maturity, quality and reliability were most important when deciding to adopt an ITS technology or system. Among the 'other' category, one participant communicated that 'degree of customization' was most important to his ITS adoption decision-making.

Participants were split among implementing organization factors when selecting which was most important. Knowledge and expertise was most important to 34 percent of the responses. Nearly another third chose "partners on board/organizational priorities" as most important. A fifth of the responses selected "adoption by peers" as the most important among organizational factors when considering adoption of an ITS technology or system.

Budget/funding sources was most often selected among webinar responses (33%) as the most important external factors during ITS adoption decision-making. A close second is agency priorities with 31 percent of responses.

Among user and market factors influencing ITS adoption decision-making, the webinar group was split as to what is most important. Responses suggest that end-user/market understanding of benefits and risk is more important to a small handful of participants compared to end-user/market acceptance and attitudes towards technology. One participant commented that another factor is the "cost of the new technology to the consumer" and continued that this can be a component of the user acceptance factor.

There is a difference in the number of factors within each category; consequently the percentages and counts among category should not be directly compared. What can be gleaned in general is that there is a strong consensus within the technological factors that demonstration of benefits followed by standards/interoperability are more important. Among other sets of factors there was no clear consensus on what is most important during ITS adoption decision-making, and rather depending on the environment of the decision, many of the factors can come into play as more important.

Factors Influencing Decision to Grow, Maintain, Contract/Cancel ITS

The webinar served as the first step in addressing a key knowledge gap identified during the Task 2 Literature Review, the lack of research on factors influencing subsequent decisions to grow, maintain, contract or cancel specific ITS technologies or systems. Whereas for the adoption decision, participants were polled on each set of factors, the time constraints of the webinar supported only a single question for the three other types of ITS decisions. Consequently, the previous four sets of factors were downsized to seven factors and an 'other' category. These seven were selected based on what researchers anticipated as being most important and include:
- Price – technology factor
- Demonstration of benefits – technology factor
- Standards/interoperability –technology factor
- Users' acceptance/attitudes –user/market factor
- Political environment –external factor
- Budget/Funding Sources –external factor
- Agency Priorities –external factor

Participants were asked to select the two factors that were most important in decisions to (1) grow, (2) maintain, and (3) contract/cancel ITS technologies or systems. However, given technical challenges in implementation and participant adherence to request, some selected a single factor while other chose two or even three factors. On average, one factor was selected by each participant when presented the decision to grow ITS technologies. This increased to an average of 1.9 factors per participant for the decision to maintaining technologies and an average of 2.2 factors per participant during the decision to contract/cancel a technology or system. The figure below present the outcomes from these three webinar poll questions.

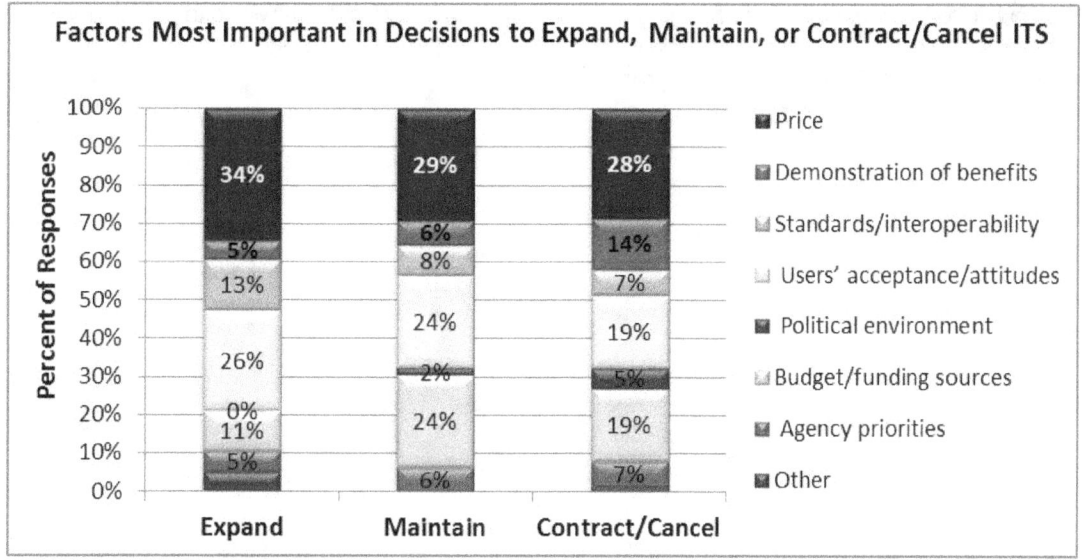

Figure E-8: Factors Selected As Most Important during Decisions to Expand, Maintain, or Contract/Cancel ITS Technologies or Systems *(Source: Noblis 2013)*

Findings of note concerning the decisions to **expand, maintain, contract or cancel** ITS projects include:
- The greatest number of participants selected "price of the technology" as the most important factor in the decision to expand, maintain, contract/cancel ITS projects.
- For decisions to expand systems, "users' acceptance/attitudes" is cited as most important by more than three times as many responses than "budget/funding sources." However, when maintaining or contracting/cancelling ITS "budget/funding sources" is cited as often as "users' acceptance/attitudes."
- Whereas the political environment was not a major factor for ITS system expansion, a few participants did cite this as the most important factor in decisions to contract or cancel ITS technologies.
- Finally, "demonstration of benefits" is a far more prominent factor when deciding to contract/cancel a system compared to decisions to adopt, expand, or maintain ITS systems or technologies.
- In the decision to contract or cancel technologies, participants submitted through the webinar chat box comments that they had not cancelled or contracted a technology, and that for one participant their ITS device was no longer supported and replacement parts were unavailable.

Webinar Conclusion and Follow up

In response to the question regarding knowledge and technology transfer products that would be helpful to next wave Integrated Corridor Management (ICM) implementers, workshops and conferences and the ICM Implementation Guide were mentioned most often, followed by webinars

Appendix E. Workshop Summary

and reliable cost-benefit information. However, due to the low number of responses, it is hard to draw conclusions regarding this question. As such, the information should be considered anecdotal.

During the webinar, all participants were invited to participle further through a 20-minute screening interview. In total, there were 22 individuals who indicated that they would be willing to participate in a follow-on interview. All of these individuals were contacted as part of the subsequent activity of the screening phone interview, the topic of Section 4.